Advanced Introduction to Water Economics and Policy

T0293201

Elgar Advanced Introductions are stimulating and thoughtful introductions to major fields in the social sciences, business and law, expertly written by the world's leading scholars. Designed to be accessible yet rigorous, they offer concise and lucid surveys of the substantive and policy issues associated with discrete subject areas.

The aims of the series are two-fold: to pinpoint essential principles of a particular field, and to offer insights that stimulate critical thinking. By distilling the vast and often technical corpus of information on the subject into a concise and meaningful form, the books serve as accessible introductions for undergraduate and graduate students coming to the subject for the first time. Importantly, they also develop well-informed, nuanced critiques of the field that will challenge and extend the understanding of advanced students, scholars and policy-makers.

For a full list of titles in the series please see the back of the book. Recent titles in the series include:

Spatial Statistics
Daniel A. Griffith and Bin Li

The Sociology of Self
Shanyang Zhao

Artificial Intelligence in
Healthcare
*Tom Davenport, John Glaser and
Elizabeth Gardner*

Central Banks and Monetary
Policy
*Jakob de Haan and Christiaan
Pattipeilohy*

Megaprojects
*Nathalie Drouin and Rodney
Turner*

Social Capital
Karen S. Cook

Elections and Voting
Ian McAllister

Youth Studies
*Howard Williamson and James E.
Côté*

Private Equity
*Paul A. Gompers and Steven N.
Kaplan*

Digital Marketing
Utpal Dholakia

Water Economics and Policy
Ariel Dinar

Advanced Introduction to

Water Economics and Policy

ARIEL DINAR

Distinguished Professor of Environmental Economics and Policy, School of Public Policy, University of California, Riverside, USA

Elgar Advanced Introductions

 Edward Elgar
PUBLISHING

Cheltenham, UK • Northampton, MA, USA

Published by
Edward Elgar Publishing Limited
The Lypiatts
15 Lansdown Road
Cheltenham
Glos GL50 2JA
UK

Edward Elgar Publishing, Inc.
William Pratt House
9 Dewey Court
Northampton
Massachusetts 01060
USA

A catalogue record for this book
is available from the British Library

Library of Congress Control Number: 2022943327

Printed on elemental chlorine free (ECF)
recycled paper containing 30% Post-Consumer Waste

ISBN 978 1 83910 958 4 (cased)
ISBN 978 1 83910 959 1 (eBook)
ISBN 978 1 83910 960 7 (paperback)

Printed and bound in the USA

To my grandchildren
Shai Matan, Ari Lev, Tali Rose, Yael Aviv
Issa Brown and Gideon Paz

Contents

Figures

Tables

About the author

Ariel Dinar is a Distinguished Professor of Environmental Economics and Policy at the School of Public Policy, University of California, Riverside (UCR). He teaches courses on water policy, global–local policy interactions, management of international water and micro-economics for public policy. His work addresses various aspects of economic and strategic behavior associated with management of water, land and the environment. Dr Dinar received his PhD from the Hebrew University of Jerusalem. Since then, he spent 15 years in the World Bank working on water and climate change economics and policy in a global context. In 2008, Dr Dinar assumed a professorship at UCR. Dr Dinar founded the Water Science and Policy Center, which he directed until 2014. Dr Dinar is a Fulbright Senior Specialist since 2003; an International Fellow of the Center for Agricultural Economic Research of the Hebrew University of Jerusalem, Israel since November 2010; and was named a 2015 Fellow of the Agricultural and Applied Economics Association. He has authored and co-authored nearly 250 publications in peer reviewed journals, policy outlets and book chapters. He has authored, co-authored and edited 32 books and textbooks. Dr Dinar founded two technical journals (*Strategic Behavior and the Environment* and *Water Economics and Policy*); for the latter one he serves at present as the Editor-in-Chief. He founded and serves as the Editor-in-Chief of the book series Global Issues in Water Policy, and serves as the Editor-in-Chief of the book series Lecture Notes in Economics and Policy.

Preface

This *Advanced Introduction to Water Economics and Policy* is the result of many years during which I practiced, researched and taught aspects of water economics and management. Several experiences have made me aware of the importance of water scarcity effects on our life and welfare, communicating economic principles of managing water and understanding the various links among components of the water sector and between the water sector and other sectors.

The first experience has to do with teaching water economics. One of the questions I always ask undergraduate students in my classes on Environmental Economics is: "Do you know where your water comes from?" The surprising answer I get year after year is: "Of course, it comes from the faucet."

My second experience took place when I was a young farmer irrigating fields of potatoes on a kibbutz farm. Potatoes, to those who do not know, are a crop sensitive to quantity and timing of irrigation. Yields of potatoes can be reduced dramatically if the farmer misses the irrigation schedule or if the quantity of water applied is lower than the prescribed amount. While irrigating a potato plot I hit a main booster of the plot causing water splash that stopped the flow of water to the sprinklers. I had to drive to the main valve of the field and shut it off. Then I had to drive to the kibbutz headquarters to call the plumbers who had to dig and replace the broken booster. By the time that the booster was fixed it was already evening. The options I had were to go back to my room and return the next day to resume irrigation or to stay at night and initiate the interrupted irrigation. At that time I was already tired, but still ran scenarios of benefit–cost analysis, including individual and social considerations and calculations (at that time in my life, not knowing yet the exact economic terms). Finally, as a conscientious young person I stayed into the night and completed the irrigation schedule. Only years later, when I had the right economic concepts, I recognized that what had happened was my realization of the opportunity cost of a decision to resume irrigation the next day. At the end of the season this plot had the highest yield of potatoes among all plots of my kibbutz. No specific reason found ...

These two experiences made me realize the importance and need to communicate economic principles to many water decision-makers and users around the world who are non-economists. Actually, the majority of water professionals who manage water and make decisions regarding water production, regulation, allocation and use in many countries are non-economists.

I hope that this book, although not comprehensive in dealing with the water sector and its issues, will still be useful in shedding light on the role of economics in water management.

Ariel Dinar
School of Public Policy
University of California, Riverside

Acknowledgement

Chapters 5–11 of this book were written, and the book was completed, while the author was on sabbatical leave, co-hosted by the School of International and Public Affairs and the Columbia Water Center at Columbia University, New York. I thank the faculty and staff of the Columbia University for their hospitality and discussions during my stay.

1 Introduction to water economics and policy

Ensuring adequate water for people and the environment is one of the greatest challenges facing humanity in the twenty-first century. The availability of water, which depends on quantity, quality and timing, influences virtually all human activities and ecosystem functions. Despite the inherent complexity of water availability, the hydrological cycle has historically been studied from narrow disciplinary perspectives, often isolated from societal context. Addressing challenges of water availability requires a holistic understanding, bringing together different disciplinary perspectives that consider the reciprocal impacts between society, the environment and water availability.

This book is organized into 11 chapters. The following describes the content of each chapter.

Chapter 2 reviews past and future trends in water availability and use. Water availability, an important aspect in economic decision-making, is driven to a large extent by population growth and by the population's ability to regulate use and pollution of this resource. Water availability trends vary across time and geographic regions, affected by natural trends and the ability to regulate water use. This chapter reviews past and future trends in water availability, indices that measure water availability, technologies that increase the supply of water, and the impact of water availability on well-being across nations. The conclusion of this chapter is the motivation for this book: the importance of economic tools and their use in addressing various water-related policy issues.

Chapter 3 focuses on irrigation water management. The irrigation sector is the biggest user of water resources, consuming on average 75–90 per cent of available water in the state. Irrigation is also used in many countries as a development policy. What are the consequences of such policy?

This chapter presents the economic questions associated with irrigated agriculture, such as charges for and pricing of irrigation water, water use efficiency, command and control policy interventions, the role of water user associations, technological effects on water efficiency, agronomic improvements and water efficiency, and more.

Chapter 4 focuses on residential water management. While the residential sector uses, on average, only 5–10 per cent of available water, it has been in the focus of extensive economic work, mainly due to the relatively easiness to obtain empirical information on household water consumption and responses to policy intervention aiming to conserve water. This chapter reviews various pricing methods in the residential water sector, residential non-pricing policies, water quality impacts on the health and well-being of households, responses of households to water conservation policies, and more.

Chapter 5 describes environment–water interactions and management. Water and environment have been the focus of many economic analyses for quite some time. Environmental amenities, not being part of market mechanisms, introduce difficulty for the economist when trying to evaluate the monetary impacts of policies and regulations on the environment. Alternative approaches have been developed to assess impacts of environmental pollution on the environment, such as contingent valuation, travel cost method, and the like.

Chapter 6 highlights economic consideration of groundwater management. Groundwater (GW) provides 40–70 per cent of the available water that is used for all sectors' consumption. Due to the fact that GW is a hidden resource, it was also far from the eye and was not properly managed and protected. As a result, many aquifers around the world are on the verge of depletion and even collapse. This chapter describes the state of GW around the world, demonstrates how economic principles are used for optimal management of GW, and covers special issues related to the interaction between GW and surface water and their possible conjunctive use.

Chapter 7 summarizes aspects of the economics of water pollution regulation. Use of water is also associated with the pollution of the environment, such as agricultural runoff containing chemicals from the agricultural processes that pollute waterways directly or indirectly. Pollution from agriculture is known to be *nonpoint source pollution* due to the difficulty in identifying pollution sources. The industrial sector is also known to pollute by disposing of polluted water that is the result of the industrial production process. Industrial pollution is known to be *point source pollution*. This chapter will discuss various economic principles and examples of policies dealing with nonpoint and point source pollution, such as pollution and technology standards, taxes on pollution, subsidies for technology adoption and trade in pollution permits.

Chapter 8 deals with economics of international water management. International water is defined as freshwater sources and fresh waterways shared by more than one country. Attention to issues related to international water has been on the rise for years, but increased scarcity and the shadow of climate change has given it a boost. Economic questions related to sharing, regulating and developing of international water will be addressed in this chapter. Background information on mechanisms developed by riparian nations to a shared water source over the past will be included as well as how various economic tools perform under various conditions.

Chapter 9 describes the interaction between climate change and water resources. One of the main threats to the water sector is climate change. While the science of climate is well established, societal readiness lags behind. The water sector is affected by climate change in two main ways: extreme events of droughts and floods. This chapter touches upon the economics of climate change mitigation and adaptation in water resources and water-consuming sectors, with a focus on the agricultural and wastewater sectors.

This book, while attempting to address important aspects of water economics and policy, could not touch upon all aspects that might be of interest to readers. Chapter 10 attempts to partially close this gap. The chapter describes three emerging issues, with economic principles for dealing with three approaches to easing water scarcity effects. The issues described include managed aquifer recharge, inter-basin and intra-basin water transfers, and intersectoral implications of use of treated wastewater for irrigation.

Chapter 11 serves as the summary and concluding remarks of the book. As can be seen from the chapters described above, water economics is complex, dealing with multiple aspects that are also interrelated. While this advanced introduction focused on sectoral and topical aspects of managing water, it is important to emphasize the need to undertake a comprehensive approach to analyzing the water economy (Dinar and Tsur, 2021). Chapter 11 attempts to aggregate the conclusions one draws from all the chapters and makes the case for the role of water economics in the sustainable management of water resources around the world.

References

Dinar, A. and Y. Tsur, 2021. *The Economics of Water Resources: A Comprehensive Approach*. Cambridge: Cambridge University Press.

2   Past and future trends in water availability and use

Many estimates and forecasts of water availability around the world do exist. These estimates and forecasts may differ in the trend they display, but no matter how they may differ, all agree that over time less and less water is available on a per capita basis. The reason for such trend is very simple. It is driven by two variables that have different trajectories: (1) annual total available renewable water resources and (2) annual population. Both variables can be measured at a region, a country, a continent or globally. While annual total available renewable water resources are more or less constant over time (except for peak flood or drought years), annual population changes over time, growing in most countries and declining in few countries. Therefore, by calculating the total available renewable water resources per capita, one obtains a hyperbolic pattern, with available renewable water resources per capita today being much lower than in the past, and available renewable water resources per capita in the future, being much lower than today. We will use two examples from two countries that represent high water scarcity and high water abundance (Morocco and Israel, respectively; see Figure 2.1).

Following Figure 2.1, what can we say about these two countries? Since 1962 Israel faced a reduction in its water availability from 330 to 90 m³/capita per year[1] (nearly 70 per cent reduction) and Morocco faced a reduction from 2200 to 850 m³/capita per year (nearly 60 per cent reduction).[2] These changes in water availability are the result of only population increase and not any other possible deterioration of water quality that affects water availability. Does the relative reduction and the absolute water scarcity values suggest that citizens of Israel face a more severe water scarcity than those of Morocco? Or that the economy of Israel is performing worse than that of Morocco? The answers to these questions are not easy to get. However, while water is indeed an important component in the economy of a state, it is not the only factor affecting its

5

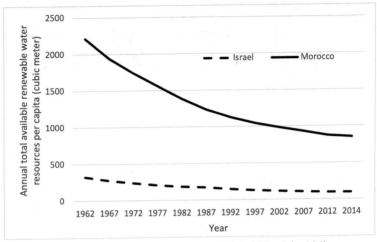

Source: Elaborated by author, based on data in World Bank (n.d.(a)).

Figure 2.1   Typical water availability changes over time in a water scarce and a relatively water abundant countries

economic performance. Institutional arrangements and technological ability are two aspects that can explain how countries that face severe water scarcity fare better in economic terms than states that have much more abundant water resources. In the case of Morocco and Israel, which should not be the rule for such assessment, population and gross domestic product (GDP) per capita (as a measure for economic wealth) in 2014 were 34.2 million and $3,100[3] per capita, and 8.3 million and $37,700 per capita, respectively.[4] The results suggest that a state with nearly one-tenth of the available water per capita (90/850) produces wealth for its economy that is more than ten times (37.7/3.2) higher.

With such opposite trends, a question that comes to mind is that maybe the total amount of available water is less important, but what makes the difference is the way such water is allocated for use in the different sectors of the economy. This is exactly what economics can help us understand. More efficient use of scarce water means development and adoption of water-saving technologies and development of products that can attract better market value either domestically or internationally. This is what we will try to do in this book through discussions and demonstrations of policies aimed at sectoral water management, economy-wide water management and internationally shared water management. But before we do that in future chapters, we still owe ourselves a discussion of historical and future water uses in the world.

2.1   Total available renewable natural water resources in the world

Available estimates of water resources in the world, or even in different geographical locations such as states, continents and globally, do not agree on how much is there. One of these estimates (Shiklomanov, 2000) suggests that the total available renewable natural water resources is 42,900 km³/year (Table 2.1). Scrutiny of the values in Table 2.1 suggests significant variations across continents and, not less important, high variation between possible low and high values of available water in each continent (that could reflect variation due to climatic conditions). We should mention that the estimates in Table 2.1 refer only to renewable natural water resources and do not take into account reuse of treated wastewater and use of desalinated water.

2.2   Changes in global water use: withdrawal and consumption by sector

There are estimates and forecasts of water use over time by sectors (Shiklomanov, 2000: Table 5) that teach us many lessons about how water was reallocated between uses, and how efficiency of water use has changed over time (Table 2.2). The values in Table 2.2 indicate an interesting dynamics of water withdrawals for use in different sectors; it also represents changes in intersectoral allocations and efficiency improvements, as we discuss. While the table includes information on the irrigated sector and on municipal use, one also has to consider the industrial sector uses and water placed in reservoirs to balance intersessional needs. We will only mention that between 1950 and 2025 (forecasted), industrial withdrawals expanded five-fold and withdrawals for storage in reservoirs increased nearly 25 times. The latter reflects responses to increased variation in water supply due to climate change.

The results in Table 2.2 suggest that world population grew three-fold during the 75 years represented in the table. In order to supply the food for the growing population, irrigated agricultural land has been expanded three-fold during that period. What about water? In order to irrigate that land, water has been withdrawn (W) and applied (A). We are interested in three indexes: (1) applied water per irrigated hectare, (2) the ratio of water applied to withdrawn and (3) the share of water withdrawn for irrigation from total water withdrawn.

All three indexes tell us important stories about changes in water use in irrigated agriculture over time. It is seen clearly that while irrigated agriculture expanded over time, applied water per hectare (A/Ha) decreases continuously over time from 7,140 to 6,840 m^3/Ha, while food production per Ha increased (Dinar et al., 2019). This decrease in applied water per hectare represents improved efficiency in water use, which is the result of technological and agronomic (genetic) innovations observed and analyzed elsewhere (e.g., Hayami and Ruttan, 1985). The second index – share of water applied to withdrawals – shows in increase from nearly 67

Table 2.1   Available renewable natural continental and world water resources, 2000

Continent	Available renewable natural water resources (km^3/year, rounded values)		
	Mean	Min	Max
Europe	2,900	2,250 (77)[a]	3,400 (117)
North America	8,000	6,900 (86)	8,900 (111)
Africa	4,050	3,100 (76)	5,100 (126)
Asia	13,500	11,800 (87)	15,000 (111)
South America	12,050	10,300 (85)	14,350 (119)
Australia and Oceania	2,400	1,900 (79)	2,900 (121)
World	42,900	39,800 (93)	44,750 (104)

Notes: Minimum and maximum columns represent possible changes in available renewable natural water resources as a consequence of wet and dry years. One cubic kilometer (km^3) is equivalent to 1 billion cubic meters (m^3) or 810,714 acre feet.
[a]   In parenthesis are percentage decrease/increase from mean value.
Source: Elaborated by author, based on data in Shiklomanov, 2000: Table 1.

Table 2.2 Estimated and forecasted world water use by sectors and time and derived efficiency (1950–2025)

Year		1950	1960	1970	1980	1990	2000	2010	2025
Population (millions)		2542	3029	3603	4410	5285	6181	7113	7877
Irrigated land (10⁶ Ha)		101	142	169	198	243	264	288	329
Water use in agriculture	W (km³/year)	1080	1481	1743	2112	2425	2605	2817	3189
	A (km³/year)	722	1005	1186	1445	1691	1834	1987	2252
	E=A/W (%)	66.8	67.8	68.0	68.4	69.7	70.4	70.5	70.6
	Applied per Ha (m³)	7140	7070	7010	7290	6950	6940	6890	6840
	Share of A to total W (%)	78	75	69	66	66	65	63	61
Municipal use	W (km³/year)	86.7	118	160	219	305	384	472	607
	Water use per person (m³/year)	34.1	38.9	44.4	49.6	57.7	62.1	66.3	77.05
Total water withdrawals (including industrial and reservoirs)	W (km³/year)	1382	1968	2526	3175	3633	3973	4431	5235

Notes: Values for 1950–1990 are estimated based on actual data; values for 2000–2025 are forecasts. W=Water withdrawals; A=Applied water; E=Efficiency measure. (1 Ha = 2.5 acres.)
Source: Elaborated by author, based on data in Shiklomanov, 2000: Table 5.

per cent in 1950 to more than 70 per cent in 2000 and beyond. This is evidence that irrigated agriculture has improved in terms of its physical efficiency – less of the extracted water is wasted. Finally, the third index tells us the story of the competition between the urban and irrigated sectors. It is clearly seen that the share of water withdrawn for irrigation as a share of overall withdrawn water for consumption in residential dwellings and industrial operations is declining over time, from 78 per cent in 1950 to 61 per cent in 2025 (estimated).

And finally, the total withdrawal of water for all uses (irrigation, municipal, industrial and reservoirs) has increased dramatically, nearly four-fold, over the period 1950–2025. Calculating the values per capita (not shown in Table 2.2 but easily calculated by dividing total water withdrawals by population) suggests that nearly 5,500 and 6,600 m^3 per capita were (and will be) withdrawn in 1950 and in 2025, respectively. These numbers indicate that society's need for larger amounts of water to be stored in reservoirs due to water supply variability following climate change effects are a main consideration in future planning of the water sector (the data on reservoirs is not shown in Table 2.2 but can be found in Shiklomanov, 2000: Table 5).

The next section provides estimates of water use in the United States during the period 1950–2015, with focus on sectors and water sources. Trends in the United States are similar to those identified in the world, although, as usual, distinction has to be made between developing and developed countries.

2.3   Example: Trends in water use in the United States, 1950–2015

It is interesting to follow, in more detail, the trends in water use in one country with a focus on sources of water and on the consuming sectors.

The case of the United States is interesting by itself since it shows a dynamic pattern of increase in withdrawals over a certain period, constant withdrawals over another period and a decline in withdrawals over the last period (Table 2.3, Figure 2.2). These trends are also characterized by changes in composition of the different sources of water withdrawn

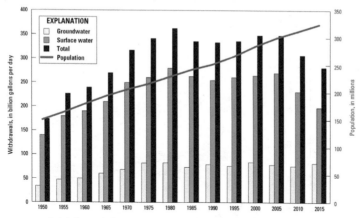

Trends in population and freshwater withdrawals by source, 1950–2015

Source: USGS (n.d.).

Figure 2.2   Trends in population changes and freshwater
withdrawals by source, 1950-2015

(USGS, n.d.). All in all, the patterns of the trends indicate effective water conservation efforts with positive effects in the past four decades.

Table 2.3 shows a couple of interesting results. First, the total water withdrawals picked up until the 1980s, including withdrawals from freshwater and saline water sources. The level of withdrawals remained then more or less flat until the beginning of the 2000s and then dropped significantly. We should also mention that population growth in the United States was linearly increasing from 1950 until 2015, and thus withdrawals per capita have declined since the 1970s. Another interesting development over the years is the increase in withdrawals for irrigated agriculture that stabled in the 1980s and then declined. In parallel it is noticeable that water withdrawals for thermoelectric power production exceeded withdrawals for irrigation in 1970, due to a hike in dam buildings after World War II, but since the 2000s the trend reversed, probably due to rising concerns from environmental damages and from a need to retire some of the existing dams.

Table 2.3   Population and water use in the United States by source and main sectors, 1950–2015

	1950	1960	1970	1980	1990	2000	2010	2015
Population (millions)	150.7	179.3	205.9	229.6	252.3	285.3	312.6	325.0
Total withdrawals (billion m³ per year)	248.6	372.9	511.0	593.8	557.9	570.3	488.9	444.7
Public supply	19.3	29.0	37.3	45.6	53.4	59.8	58.0	53.8
Self-supplied domestic	2.9	2.8	3.6	4.7	4.7	4.9	4.9	4.5
Irrigation	122.9	151.9	179.5	207.1	185.0	191.9	160.2	162.9
Thermoelectric	55.2	138.1	234.8	290.0	267.9	269.3	223.7	183.7
Industrial	51.1	52.5	64.9	62.1	30.9	26.9	22.4	20.4
Mining	(a)	(a)	(a)	(a)	6.8	5.7	5.5	5.5
Aquaculture	(a)	(a)	(a)	(a)	3.1	8.0	12.4	10.4
Total groundwater[b]	46.9	69.6	95.3	115.9	111.4	119.8	107.8	116.9
Total surface water[b]	207.1	305.2	418.4	457.1	447.0	450.2	381.1	326.7

Notes: [a] Included in industrial; [b] Including fresh and saline.
Source: USGS (n.d.) Original values are expressed in billion gallons per day. (1 billion gallons per day = 1.381 billion m³ per year.)

2.4   Summary and the plan ahead

In this chapter we reviewed the changes in water withdrawals and consumptions over time and among consuming sectors. We realized that in the past 70 years or so, there have been changes in patterns of consumption that were the result of preferences of water users, technological changes and policy impacts. The view we have so far is a global one, with an example from one big country: the United States. We know, so far, that water becomes increasingly scarce, that success of the regulations of its use depends on many factors, including political, economic and social, to name a few.

In the next chapters we will delve into the economic principles that are the basis of regulating the use and optimal allocation of water among user sectors, with a focus on the irrigation sector and the residential sector, both of which consume the lion's share of the water on Earth. We will also discuss the regulation of different types of water, including surface water and groundwater.

Notes

1. $1235 \text{ m}^3 = 1$ acre-foot.
2.   Calculated as available renewable water resources/population size.
3.   All monetary values used in this book are USD at the year they are cited.
4.   Source for population and GDP per capita are World Bank (n.d.(b)).

References

Dinar, A., A. Tieu, and H. Huynh, 2019. Water scarcity impacts on global food production. *Global Food Security*, 23:212–26.

Hayami, Y., and V. W. Ruttan, 1985. *Agricultural Development: An International Perspective*. Baltimore: Johns Hopkins University Press.

Shiklomanov, I. A., 2000. Appraisal and assessment of world water resources. *Water International*, 25(1):11–32.

USGS (United States Geological Survey), n.d. Trends in water use in the United States, 1950 to 2015. Accessed 25 February, 2022. https://www.usgs.gov/special -topic/water-science-school/science/trends-water-use-united-states-1950 -2015?qt-science_center_objects=0#qt-science_center_objects.

World Bank, n.d.(a). Renewable internal freshwater resources per capita (cubic meters), presenting from Food and Agriculture Organization, AQUASTAT data, 1962–2014. Accessed 25 February, 2022. https://data.worldbank.org/indicator/ER.H2O.INTR.PC?end=2014&start=1962&view=chart.

World Bank, n.d.(b). World development indicators. Accessed 25 February, 2022. http://datatopics.worldbank.org/world-development-indicators/.

3 Management of water in the agricultural sector

Agriculture is the largest user of water on Earth. Globally, more than 70 per cent of available freshwater is used in agriculture (World Bank Blog, 2017). However, major variations are seen across regions (Figure 3.1) and individual countries.

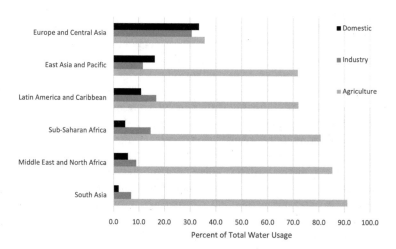

Source: Elaborated by the author, based on data in World Bank Blog (2017).

Figure 3.1   Per cent use of freshwater by major sectors in regions of the World Bank

The share of water used in irrigated agriculture is also a function of the level of development of a country. The less developed a country is, the more it relies on agriculture and thus the greater is the share of its water resources used for irrigation. For example, share of water abstracted for irrigation in 2015 was 15 percent in Kenya (a developing country) and 1 per cent in the Netherlands (a developed country), while the share of agricultural GDP was 34 per cent and 1 per cent, respectively (World Bank, n.d.).

Because of its central role in both developing and developed countries, water resources are the focus of many intervention policies (Dinar et al., 1997; Dinar and Mody, 2004; Tsur et al., 2004). These policies have been aimed at achieving multiple objectives, including income transfer, food production security, environmental sustainability and resource conservation. Since agriculture consumes the largest share (70–90 per cent in most countries) of annual renewable freshwater on earth it is a sufficient reason for policy makers to focus major efforts on improved performance of scarce water use in irrigated agriculture.

The most important policy objectives for irrigated agriculture are food security and efficient water use. Governments intervene in order to achieve these objectives by introducing several policy tools such as water pricing (or charges), water quotas, water subsidies and several other policy interventions, such as the introduction of water markets and empowerment of water user associations, which will not be discussed in this chapter. Additional government interventions aimed to achieve higher water use efficiency include provision of know-how by making public extension services available to farmers (Chatterjee et al., 2019), and providing weather information needed for irrigation decision-making (Parker and Zilberman, 1996). These last two interventions are achieved by creating a public good – the information and know-how that are provided for free to water users in the irrigated agricultural sector.[1]

In recent decades policy makers became increasingly aware of the negative externalities associated with water use in agriculture. In particular, two types of negative externalities have been recognized and addressed: (1) groundwater pumping by farmers for irrigation purposes that leads in many cases to lowering of water level in the aquifers, which leads to higher pumping cost to all aquifer users, and (2) water quality deterioration in water bodies due to disposal of irrigation water deep percolation laden with chemicals from the agricultural process into rivers and groundwater aquifers.

3.1   Water use efficiency

One objective in addressing growing water scarcity is to increase water use efficiency (WUE) across different irrigators. Put it simply, water use efficiency in an irrigation system refers to the ratio between the actual water volume consumed at the crop root zone to the total water volume entered into the main delivery system. While irrigated agriculture contributes to the objective of food security, WUE in the irrigation sector is generally very low in many countries, averaging 25–50 per cent (Tiwari and Dinar, 2002). Increase in WUE in irrigated agriculture alone could meet about 50 per cent of the projected increase in total water consumption, with water conservation in agriculture being considered as a new 'source' of water (Caswell, 1991).

But WUE has several faces. Engineers use the concept of *technical efficiency* of irrigation water use, usually measured as the ratio between water supplied by the system to water taken by the plant during the irrigation season. Economists refer to *economic efficiency*, measured in terms of crop output (mainly in monetary value) per unit of water applied, or overall financial returns in terms of net benefits from the irrigation project per unit of water delivered. Here we may face situations where a distinction is made between private-level and social (basin)-level efficiency. This brings us to another definition – *ecological or environmental efficiency*, which introduces water-dependent ecosystems beyond the agricultural crops (Esteban and Dinar, 2016). Therefore, water should be allocated such that it meets the need for consumptive use of water without adversely affecting the ecological health of surroundings uses, beyond the irrigation project, basin-wide, etc. (Ward and Pulido-Velazquez, 2008).

Once we have a measure of WUE we can then compare different policy interventions and prioritize among them, based on the criteria of WUE we selected. But the irrigation sector has several characteristics that make its water management complicated. Since developing and developed countries differ in their institutional capacity and level of wealth, which translates into different infrastructure levels, the way that water use in agriculture is regulated must be different as applied to a developed or developing country. These characteristics will be highlighted in the discussion below, where the different policy interventions are addressed in sections 3.2–3.4.

Table 3.1   Irrigation water pricing methods, application principles and challenges

Pricing/charging method	Application principles	Advantages and challenges
Volumetric (single rate unified)	Irrigators pay a given rate per volume of water ($/m³) they consume. The method necessitates a water measuring device. Variation of the volumetric approach include (1) indirect calculation based on measurement of flow time (as from a reservoir) or time of uncertain flow (proportions of a flow of a river), and (2) a charge for a given minimal volume to be paid for even if not consumed (to secure the providing agency's revenue).	The unified rate is easy to implement and enforce, once the measuring devise functions well. It sends a clear message to the users who are able to make decisions facing one single rate. However, since the rate is the same for the entire range of the volume consumed, the incentives to save water by users are not that effective.
Volumetric (inclined tiers)	Irrigators pay per volume based on the volume consumed $\left(\dfrac{\$_1}{m_1^3} > \dfrac{\$_2}{m_2^3} > ...., > \dfrac{\$_n}{m_n^3}\right)$. This method necessitates a water measuring devise as well. This is a multi-rate volumetric method, in which water rates vary as the amount of water consumed exceeds certain threshold values that are set by the water authority. Usually rates are higher for larger volumes. Number of tiers could be greater than two.	The inclined tier method is considered to be the most effective pricing method in portraying the scarcity value to the irrigators. It is also complicated for users to follow and to the water agency to administer. It is associated in many locations with political disagreements and negative and destructive behavior on the part of the irrigators.

Pricing/charging method	Application principles	Advantages and challenges
Per unit area	Irrigators are charged per unit of irrigated area, depending on the kind and extent of the crop irrigated, irrigation method, the season of the year, etc. Pumped water is usually charged higher than gravity water. Farmers are required to pay and, in some cases, the per area charges are also for non-irrigated land.	The per unit area pricing method is the most popular charging (not pricing) method for irrigation water, especially in developing countries where water supply from irrigation projects is not functioning well (reliability is questionable). This method is easy to administer. Charging per unit area is meant mainly to recover the investment cost of the irrigation project rather than to increase efficiency. In many cases charging per unit area doesn't provide any incentive to irrigators to become more efficient.
Output pricing	Irrigators pay a water fee per each unit of the output they sell. No water-measuring device is needed.	Output pricing is a simple water charging method that aims mainly to recover the investment cost. It is easy to monitor, especially if the product is sold in official (government) outlets.
Input pricing	Irrigators pay a water fee per each unit of a certain input bought for use in the production process. No water-measuring device is needed.	Input pricing is usually associated with regulation of pollution from use of chemicals (fertilizers or pesticides) that contaminate water bodies.

Pricing/charging method	Application principles	Advantages and challenges
Two-part water tariff	Involves charging irrigators a constant marginal price per unit of water purchased (volumetric marginal cost pricing) and a fixed annual (or admission) charge for the right to purchase the water. A water-measuring device is needed. The admission charge is the same for all farmers. This pricing method has been advocated, and practiced, where a public utility produces with marginal cost below average cost and must cover total costs (variable and fixed).	Two-part water tariff is aimed mainly to help the water utility cover the fixed costs of operation and not for improved water efficiency. Having the two parts in the tariff makes it harder to administer and the level of the fixed payments is a subject for political disputes between the users and the utility.
Betterment levy	Water fees are charged per unit area, based on the increase in land value accruing from the provision of irrigation water.	Similar to the per unit area method, but is subject to a more variation over regions due to the added value from having available water to the specific location.

Source: Adapted from Dinar and Tsur (2021).

3.2   Water pricing/charges

Water pricing is one of the main policy interventions used to regulate water use, leading to increased efficiency. Several objectives can be identified. In some countries, water pricing (charges) aims to achieve recovery of the investment in the irrigation project. In other countries, water pricing aims to achieve increased efficiency by highlighting the scarcity value of the irrigation water. It is obvious that for these two objectives, different irrigation water pricing methods need to be applied.

The reader can find more details and comparisons on the various irrigation water pricing methods in Table 3.1. More detailed information can be found in Bosworth et al. (2002).

3.3   Water quotas

Restricting the quantity of water that is delivered to irrigators could be an effective policy intervention that incentivizes irrigators to save water and become more efficient. Irrigators facing water quotas may be prompted to reconsider the cropping mix they grow, invest in irrigation technologies that deliver water more efficiently to their fields, and to implement more water-saving management practices.

Water quotas can be specific by crops (taking into account the crop needs), by geographical regions (taking into account the regional climate), and by size of the farm (taking into account the relative efficiency, or inefficiency, of different sizes of farms). Water quotas are imposed on an annual basis, but can also be imposed on a monthly basis. It is not rare to see a certain water quota in summer months (where water in the region is the scarcest) alongside an annual water quota. In such a quota arrangement it could well be that some of the quotas, say in certain months, will be more effective than in other months, or that the annual quota will not be effective in limiting the individual farmer.

While the quota system is an effective policy intervention, the transaction costs of implementing it are very high. It is usually a government agency that calculates the quotas and assigns them to regions and individual farmers. It is also quite typical that the quota system is subject to many disagreements among the farmers and the regulatory agency and leads to political disputes.

3.4   Water-related subsidies

Subsidies usually have a bad connotation and are seen as an inefficient policy intervention. However, water regulation agencies have attempted at providing special programs that include government subsidized loans to upgrade existing irrigation technologies to more efficient ones (Dinar and Yaron, 1992). Dinar and Yaron (1992) found that a subsidy rate for irrigation equipment of 20 per cent would result in reaching the ceiling (the highest level of use among users) four years earlier than with a subsidy rate of 15 per cent, whereas with a subsidy rate of 25 per cent, the diffusion reaches ceiling seven years earlier. The time gained by the increased subsidy can be weighed against possible social losses resulting from continued use of the existing inefficient irrigation technology. This brings us to the discussion of substitution between policy interventions to achieve a policy goal.

Dinar and Yaron (1992) estimated the substitution between two water policies – water pricing and water subsidy – in order to achieve the ceiling of the diffusion process for drip irrigation in a given year (they used the year 2000 as their target). They found that the substitution rate is larger in cases where water prices are high (substitution rate of 0.1) than in cases of low water prices (substitution rate of 0.05). These findings can, therefore, help the policy maker optimize the combination of these two policies, if the policy maker decides to consider a combination of these policy tools.

3.5   Packaging and sequencing of policy interventions

Esteban and Dinar (2013) offer the concept of packaging and sequencing of policy interventions (instruments) for groundwater regulation, using a hydro-economic model that was developed for the Western la Mancha aquifer in southeast Spain. Several policy interventions are evaluated separately and jointly (and in sequential manner). The main conclusion is that when policies are packaged and sequenced with triggers that dictate their implementation, there is an increase in efficiency (higher benefits, welfare and water stock in the aquifer) compared with the policies implemented separately.

In particular, when policies are implemented individually, water taxes (both uniform and differential by crop types) are preferred to a quota. In the case of water taxes, a differential tax outperforms a uniform tax. A differential tax is not only more equitable (because the polluter pays the principle) but also more efficient in terms of social welfare and farmers' gross margin. The differential tax instruments are associated with complicated application, in terms of setting the tax level and information needs and administrative costs.

The results of the sequential package of quota and tax outperform individual policy instruments. Sequential packaging obtained both higher net present value of private profits and higher values of net present value of total social welfare. There is clearly an increase in efficiency with the sequential package of quota and tax compared with the implementation of individual policy instruments. We should also provide some words of caution. While a sequential package is the efficient policy instrument, one has to observe that it is the longest to reach a steady state in the aquifer water table, and the least stable along the planning horizon path. These points are critical in designing policy reforms and could lead to opposition on the part of the farmers and likely failure of the policy (Mukherji and Shah, 2005).

Note

1. A public good is a commodity or service that is made available to all members of a society. The consumption of a public good by one member of the society will not exclude others from consuming it as well. Usually, these commodities or services are managed by governments and funded collectively through taxation.

References

Bosworth, B., G. Cornish, C. Perry, and F. van Steenbergen, 2002. *Water Charging in Irrigated Agriculture: Lessons from the Literature.* Report OD 145, December, HR Wallingford.

Caswell, M. F., 1991. Irrigation technology adoption decisions, in A. Dinar and D. Zilberman (eds.), *The Economics and Management of Water and Drainage in Agriculture,* 295–312. Boston: Kluwer Academic Press.

Chatterjee, D., A. Dinar, and G. González-Rivera, 2019. Impact of agricultural extension on irrigated agriculture production and water use in California. *Journal of the American Society of Farm Managers and Rural Appraisers,* 1:65–84.

Dinar, A. and J. Mody, 2004. Irrigation water management policies: Pricing and allocation principles and implementation experiences. *Natural Resources Forum,* 28:112–22.

Dinar, A. and Y. Tsur, 2021. *The Economics of Water Resources: A Comprehensive Approach.* Cambridge: Cambridge University Press.

Dinar, A. and D. Yaron, 1992. Adoption and abandonment of irrigation technologies. *Agricultural Economics,* 6(4):315–32.

Dinar, A., M. W. Rosegrant, and R. Meinzen-Dick, 1997. *Water Allocation Mechanisms: Principles and Examples.* World Bank Policy Research Working Paper 1779, Washington, DC.

Esteban, E. and A. Dinar, 2013. Modeling sustainable groundwater management: Packaging and sequencing of policy interventions. *Journal of Environmental Management,* 119:93–102.

Esteban, E. and A. Dinar, 2016. The role of groundwater-dependent ecosystems in groundwater management. *Natural Resource Modeling,* 29(1):98–129.

Mukherji, A. and T. Shah, 2005. Groundwater socio-ecology and governance: A review of institutions and polices in selected countries. *Hydrogeology Journal,* 13:328–45.

Parker, D. D. and D. Zilberman, 1996. The use of information services: The case of CIMIS, *Agribusiness,* 12(3):209–18.

Tiwari, D. and A. Dinar, 2002. Balancing future food demand and water supply: The role of economic incentives in irrigated agriculture. *Quarterly Journal of International Agriculture,* 41(1/2):77–97.

Tsur, Y., A. Dinar, R. Doukkali, and T. Roe, 2004. Irrigation water pricing: Policy implications based on international comparison. *Environment and Development Economics,* 9(6):735–55.

Ward, F. A. and M. Pulido-Velazquez, 2008. Water conservation in irrigation can increase water use. *PNAS,* 105(47):18215–20.

World Bank Blog, 2017. Globally, 70% of freshwater is used for agriculture. Accessed 25 February, 2022. https://blogs.worldbank.org/opendata/chart-globally-70-freshwater-used-agriculture#:~:text=In%20most%20regions%20of%20the,percent%20increase%20in%20water%20withdrawals.

World Bank, n.d. World development reports 1960–2019. Accessed 25 February, 2022. https://data.worldbank.org/indicator/NV.AGR.TOTL.ZS.

4 Management of water in the residential sector

While the residential sector consumes on average between 8–10 per cent of available water (see Figure 3.1), the attention and number of studies published on residential water management is significant. Household water consumption regulations have been the subject of many policy studies for several reasons. Household water consumption is relatively easy to monitor and regulate, and since household water consumption has the highest priority in water allocation, any conservation, even small, in residential water consumption will affect and benefit the remaining water consuming sectors that compete over scarce water resources in the economy.

4.1 The nature of household water consumption

Water is used in residential dwellings for indoor and outdoor purposes. Distinction should be made between single-family homes and apartment buildings when it comes to outdoor uses. Indoor end uses as percentage of total household consumption include (list and shares typical to developing countries): toilet (24 per cent), faucet (20 per cent for cooking and teeth brushing); shower (20 per cent); clothes washer (16 per cent); leaks (13 per cent); bath (3 per cent); dishwasher (2 per cent); and other (3 per cent) (DeOreo et al., 2016). Volume of water varies by number of household members and thus is not presented here.

Residential indoor water use is subject to considerable variation across households depending on the size and composition (adults and children) of the family. Significant reductions in some of the end uses of water listed above could be achieved not only through the adoption of efficient technologies (such as fixtures and appliances) but also through behavioral changes of users aimed at reducing water use and wastage (Suárez-Varela and Dinar, 2020) and by eliminating customer side leakage through automated metering and leak alert programs (Firat et al., 2021).

Efforts to regulate and save indoor water focus mainly on sending price signals that promote savings, on education of users to shut faucets when performing certain hygiene activities (teeth brushing) and on rebate programs to replace water-conserving toilets and showerheads.

The outdoor residential water uses in single homes (mainly in developed countries) include landscape irrigation, filling and back-washing swimming pools, washing pavement and cars and other outdoor uses. Annual outdoor use differs by climatic region. The average outdoor use across nine sampled cities in a 2016 study by DeOreo et al. (2016) was 191 m^3 per household per year. Nearly 20 per cent of homes in the DeOreo et al. (2016) study irrigated their landscapes in excess of theoretical irrigation requirement. Conserving of outdoor water will have the highest impact on conservation and is the target of many policy interventions. It is estimated (EPA, n.d.) that on average, 30–50 per cent of household water consumption is in outdoor uses, depending on the climate region.

4.2   How water consumption of households is regulated

Referring to single family homes, a regulatory agency in charge of water use can promote conservation by using incentives both for indoor and outdoor water consumption, as described below.

4.2.1   Limits on water consumption in the residential sector

A traditional regulatory intervention that is used quite frequently is a restriction on the number of weekly landscape watering days, imposed by the water supplier in a given part of the city. It is enforced usually through a schedule indicating the number of days per week a specific neighborhood landscape irrigation is permitted. Once this is set, the water utility fixes up a schedule such that not all the homes in the neighborhood are permitted to irrigate on the same day. The day restrictions are applicable only to spray systems and not to hand watering or drip and micro-sprinkler irrigation.

Hayden and Tsvetanov (2019) analyze the impact of such policy in California during the 2012–2017 drought. Findings suggest that number of restriction days have different impacts. One-, two-, three- and four-day restriction policies are not statistically different from each other and from the 2013 levels. Five- and six-day restriction per week are the only policy interventions that were found statistically significant, with water savings compared with the 2013 level of 3.9 and 9.3 per cent, respectively. The seven-day irrigation ban was also not significant, probably due to (permitted) over-irrigation with hand watering once sprinkler irrigation was not permitted.

Other regulatory restrictions (that will not be further discussed in this chapter) include: bans on using residentially supplied water for cleaning of driveways, sidewalks or vehicles (for vehicle cleaning, some specifications are provided); irrigating outdoor landscape such that runoff flows into neighboring properties and the street; and using non-recirculated potables water for decorative fountains (Hayden and Tsvetanov, 2019).

4.2.2   Rebate programs for water conserving fixtures and landscape

Households are also incentivized through a rebate program to conserve water by replacing less efficient water-using appliances with more efficient ones. The following list is taken from the San Bernardino Municipal Water Department in California (https://www.sbmwd.org/). Their indoor program includes (partial list): (1) A rebate of up to $100 with the purchase and installation of high-efficiency toilets that use 1.28 gallons per flush or less; (2) a $20 rebate for the purchase and installation of a low-flow showerhead; (3) a $100 rebate for the purchase and installation of a high-efficiency washing machine with a Consortium for Energy Efficiency (CEE) rating of Tier 1 or greater; (4) a $75 rebate for the purchase and installation of a single high-efficiency dishwasher with a CEE Rating of Tier 1 or greater; (5) a free household water conservation kit; and (6) a free easy-to-install kitchen aerator, two bathroom aerators, a shower timer and two leak detection dye tabs, leading to substantial water savings over time.

The outdoor program includes (partial list): (1) an artificial turf rebate of $2 per square foot (up to $400) for replacing irrigated grass turf with artificial turf; (2) a 50 per cent rebate (up to $150) for installing a drip irrigation system; (3) a 50 per cent rebate (up to $300) for introducing drought-tolerant trees, plants and shrubs into the landscaping; and (4) a $250 rebate for installing a weather-based irrigation controller.

4.2.3   Education and campaigns to conserve water in the residential sector

Studies show that education and mind-change campaigns are cost effective strategies for water savings. Strategies may include (partial list): (1) free water-smart landscape workshops to help reduce water use without sacrificing the outdoor beauty of the home; (2) advocacy and influencing leaders (e.g., political advocacy, religious leaders, other persons of influence); (3) school programs; and (4) involving young people as promoters and activists (Schaap and van Steenbergen, 2001).

4.2.4   Pricing as a policy intervention to conserve water in the residential sector

In many (especially developing) countries the objective of charging for water is to recover the government investment and fixed costs of water projects (this holds also for the irrigated agricultural sector – see Table 3.1, charging per unit area). By charging households periodical fixed fees (Table 4.1), the government can accumulate revenue to cover the investment and the cost of water provision to the community. But fixed fees per household doesn't send the water users a signal of the severeness of water scarcity and, thus, may not lead to conservation. While Table 4.1 includes information on several water pricing schemes, we discuss in the following text the inclined block rate (IBR) and the water budget rate (WBR). Both IBR and WBR are designed to increase household conservation efforts via realization of penalties due to wastage. Both IBR and WBR take into account the marginal cost of providing additional quantities of water from sources that are increasingly expensive and reflect the scarcity value in the pricing scheme. Therefore, both IBR and WBR charge the users higher prices for consumption of larger quantities. The size of the blocks is subject to local consideration of the individual utility, including political considerations. The usual number of blocks varies between three and five.

While IBR and WBR aim at the same objective, IBR doesn't differentiate between different sizes of households or special needs of different families. These aspects are taken into account by the WBR, which provides indoor allowances for consumption to families of different sizes (a very considerate price per unit of water in Block 1), and special considerations for families with outdoor special circumstances (e.g., horses, fruit trees), using a relatively reasonable price per unit of water in Block 2. Beyond Block 2, consumption is charged with highly exceptional prices per unit of water in Block 3, or sometimes even in Block 4 or Block 5. More information on the design of WBR and its implementation can be found in Barr and Ash (2015).

4.3   Water conservation policies in California's residential sector

With a highly water-dependent economy and precipitation patterns that vary geographically and within and across seasons, the state of California's water sector is very vulnerable to external climatic shocks (climate-induced droughts) as well as changes in demands by an ever-growing population and dynamic agricultural sector. California has experienced many droughts, including in 1841, 1864, 1924, 1928–1935, 1947–1950, 1959–1960, 1976–1977, 1986–1992, 2006–2009, 2011–2019 and 2020–2021. In recent years droughts in California have become longer and harsher, making allocation decisions about water among sectors and across locations harder.

Continuous unequal droughts conditions across the state in recent years, spanning from 1976 through 2020 (Figure 4.1) have led the California State Legislature to introduce various policies related to water conservation, declaring conservation a 'way of life' (see Table 4.2 for major policies/regulations enacted during 1976–2016).

Table 4.2 suggests a mix of interventions, including supply augmentation and top-down demand-side management. The demand-side management policies fostered a mix of water pricing initiatives and additional time of use regulations or programs for water technology upgrades. For example, the Water Conservation Act (SB X7-7), enacted at the end of the 2007–2009 drought period, mandated all water suppliers to increase their

Table 4.1   Residential water pricing methods, application principles and challenges

Pricing/ charging method	Application principles	Advantages and challenges
Household fixed fee (FF)	A fixed fee charged to the household periodically (monthly or bi-monthly) no matter how much water the household consumed.	Easy to administer. But doesn't send any signal about scarcity to the user and thus is not considered a water-saving regulation.
Flat/ uniform fee (volumetric)	The household is charged a fixed fee per unit of water consumed no matter how much water was consumed.	Uniform fee sends the signal of scarcity, depending on the level of the fee, but is not efficient enough due to having one fee per unit of water consumed along the entire consumption range.
Inclining block rate (IBR) (volumetric)	The household is charged, usually, the marginal cost, in blocks, reflecting segments of consumption, with higher ranges of consumption facing higher rates per unit of water consumed. Number of blocks can reach five.	IBR is considered an efficient pricing scheme because of the increasing rate faced by users at higher consumption levels. Could be consider a non-equitable pricing scheme because it is not taking into account the family size.
Declining block rate (DBR) (volumetric)	The household is charged in blocks, with higher ranges of consumption facing lower rates per unit of water consumed.	DBR is not a water saving scheme. It encourages consumers to consume more water and thus cover better fixed costs of the utility.

Pricing/ charging method	Application principles	Advantages and challenges
Water budget rate structure (WBRS) (volumetric)	Similar to IBR except that the first block is flexible and subject to adjustment (volume), based on special circumstances such as family size, livestock on premise, etc. First block is usually relatively very low in rate per unit of water. Remaining blocks (3–5) are very high.	WBRS adjusts the length of the first block to the needs of the household. As such, it is expected to be more acceptable and demonstrate highest effectiveness among all water conservation (volumetric) pricing methods.
Note about apartment buildings	In the case of apartment buildings with metering devices per family, the above methods apply to each family in the apartment building. In addition, the difference between consumption of the apartment building community, measured in the apartment building main meter, and the sum of the consumption of all apartments' meters (water usually used for the common areas) is divided according to agreed-upon allocation rules among all families. In the case that there are no meters for individual apartments, the entire water bill is divided among the families according to an agreed upon allocation rule (by apartment size, family size, etc.).	

Note: In most pricing methods households are charged with an additional fixed cost such as 'connection fee' paid by the household upon establishing the contract with the water utility. The rest of this chapter will focus on specific examples, using the case of California, which can be seen a leader in development of water conservation policies, due to the extreme water scarcity the state faces.

Source: Adapted from Dinar and Tsur (2021).

water use efficiency with a target of 20 per cent reduction in per capita water consumption by 2020. It was left for the water suppliers (utilities) to select the means by which such reduction could be achieved. Another severe drought in California ended in 2017. During this drought, in 2015, the governor imposed a mandatory conservation decree that required the public water agencies to reduce water consumption between 4 per cent to 36 per cent, depending on their daily per capita consumption levels in 2013, in order to achieve an overall reduction of 25 per cent. While many utilities achieved these reductions, by mid-2018 water use in many locations had rebounded to pre-drought levels. Thus, many conservation measures had temporary effects and this is recognized by the regulator. Since the historic drought of 2011–2017, California is revising its approach to water conservation. The state will no longer issue mandates for all utilities to achieve the same per cent reductions in usage.

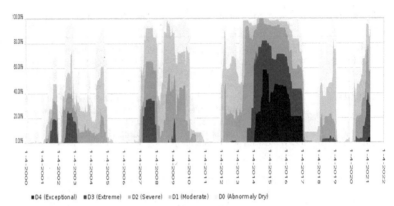

■D4 (Exceptional) ■D3 (Extreme) ■D2 (Severe) ▫D1 (Moderate) ▫D0 (Abnormaly Dry)

Source: US Drought Monitor (https://droughtmonitor.unl.edu/Data/Timeseries .aspx).

Figure 4.1   Drought severity (height of D2, D3, D4) and extent (time duration of D2, D3, D4) in California, 2000–2020

Such uniform orders do not account for differences in past conservation achievements and customer characteristics. Two statewide water conservation bills, passed in 2018, aim to make conservation a way of life. This new legislation intends to reduce per capita use by requiring utilities to meet targets for indoor and outdoor use. Water utilities have flexibility in deciding how to meet these targets.

The performance of the various conservation regulations by hydrologic regions of the state (Lee et al. 2021) suggests that the effectiveness of the regulations varies across regions and that the bouncing effect after the drought is significant (with variations by hydrologic regions). However, all regions exhibit reductions in consumption based on gallons per capita per day (GPCD), with the Central Coast region showing a reduction from 230 to 105 GPCD (54 per cent) and the San Francisco Bay region showing a reduction from 110 to 75 GPCD (32 per cent) between 1994 and 2019. These differences in conservation effectiveness across regions in the state suggest that both physical conditions as well as population responsiveness and water utility decisions and actions play a role in reaching the conservation target.

Table 4.2   Major water conservation regulations in California 1976–2016

Term [legal code]	Drought period	Enactment Year	Description
Urban Water Management Planning (UWMP) Act [AB797]	1976–1977	1983	Certain urban water suppliers produce a water management plan at least once every five years.
Amended the UWMP Act of 1983 [AB 11X and AB 1869]	1987–1992	1991	Developing water shortage contingency plans and submit them to the State under AB 11X and including requirements for water conservation and recycling under AB 1869.
Memorandum of Understanding Regarding Urban Water (MOU) [AB 892]	1986–1992	1993	Incorporating specific urban 'Best Management Practices (BMPs)' to guide water suppliers to design comprehensive conservation programs based on sound economic criteria and water conservation on an equal basis with other water management options.
Water Conservation Act [SB X7-7]	2007–2009	2009	Mandating the State to achieve a 20% reduction in urban per capita water use by 2020 relative to a 10-year historical baseline by requiring all water suppliers to increase their water use efficiency.
Law for Reporting Water Loss [SB555]	2011–2016	2015	Mandating the State Water Board to develop water loss performance standards for urban retail water suppliers and provide technical assistance to guide their water loss programs, and submitting a validated water loss audit annually to the California Department of Water Resources.
Water Conservation and Drought Planning Legislation [AB 1668 and SB 606]	2011–2016	2018	Requiring the state's ongoing efforts to make water conservation a way of life in California by provisions for efficient water use and a framework for the implementation and oversight of new standards by 2022 by water suppliers.

Note: *AB=Assembly Bill; SB=Senate Bill.
Source: Adapted from Lee et al. (2021).

4.3.1   Certain supply-side approaches

Supply-side approaches are known for their development of freshwater resources, such as dams and/or conveyance systems that move water between regions (Purvis and Dinar, 2019). However, policies to conserve water in California allow the water suppliers to take actions leading to the reduction of freshwater consumption by urban users. This has led urban water utilities to invest in water treatment technologies that (1) significantly increase the share of household sewage that is treated and reused and (2) artificially store in aquifers different types of water in abundant water years for use in drought years.

Treating municipal sewage has seen a significant increase around the world in the past 25–30 years. The number of facilities has risen from 18,000 to 72,000 and global capacity has grown from 255,000 to 307,000 billion m^3/year. The potential of reuse for irrigation has not yet been reached. For example, while wastewater treatment to the highest level of water quality is a regulatory requirement in California, most of the treated municipal wastewater is disposed of to waterways, including the Pacific Ocean. While reuse of treated wastewater in California between 2001 and 2015 has risen from nearly 650 to 950 million m^3/year (Dinar and Tsur, 2021: Table 1.4), this is not more than 2–3 per cent of the total water use in the state. The majority of the reused quantity is for landscape irrigation in urban areas (17–18 per cent) and agricultural irrigation (31–45 per cent). As suggested by Reznik and Dinar (2020), nearly 1.5 billion m^3 of good-quality, treated municipal wastewater was discharged directly into California coastal waters in 2015. Regional arrangements for use of urban treated wastewater in irrigated agriculture might be, under certain conditions, a preferred social solution to local water scarcity.

A second supply-side approach practiced in several locations around the world, including California, is managed aquifer recharge (MAR). MAR is a set of practices and institutions that allows the recharge of water of various types and qualities (surface water, recycled wastewater and even groundwater from different locations) into a given aquifer (Luxem, 2017; Reznik et al., 2021: 5). MAR is used by both agricultural and urban water agencies to bridge periods of uncertain water supplies. In the urban sector of California, the most known example is the Orange County Water District, which uses MAR mainly to intercept seawater intrusion from destroying the water quality of aquifers.

References

Barr, T. and T. Ash, 2015. Sustainable water rate design at the western municipal water district: The art of revenue recovery, in A. Dinar,V. Pochat, and J. Albiac-Murillo (eds.), *Water Pricing Experiences and Innovations*, 373–92. Dordrecht: Springer.

DeOreo, W. B., P. Mayer, B. Dziegielewski, and J. Kiefer, 2016. *Residential End Uses of Water, Version 2: Executive Report*. Water Research Foundation.

Dinar, A. and Y. Tsur, 2021. *Economics of Water Resources: A Comprehensive Approach*, Cambridge: Cambridge University Press.

EPA (US Environmental Protection Agency), n.d. *Outdoor Water Use in the United States*. Accessed 3 January, 2021. https://19january2017snapshot.epa .gov/www3/watersense/pubs/outdoor.html.

Firat, M., S. Yilmaz, A. Ateş and Ö. Özdemir, 2021. Determination of economic leakage level with optimization algorithm in water distribution systems. *Water Economics and Policy*, 7(3): 2150014.

Hayden H., and T. Tsvetanov, 2019. The effectiveness of urban irrigation day restrictions in California. *Water Economics and Policy*, 5(3):195001.

Lee, J., M. Nemati, and A. Dinar, 2021. Historical trends of residential water use in California: Effects of droughts and conservation policies. *Applied Economic Perspectives & Policy*, 44(1):511–30.

Luxem, K., 2017. Managed aquifer recharge in California: Four examples of managed groundwater replenishment across the state. American Geosciences Institute, 29 September, 2017. Accessed on 25 February, 2022. https://www .americangeosciences.org/sites/default/files/CI_CaseStudy_2017_2_MAR _170925.pdf.

Purvis, L. and A. Dinar, 2019. Are intra- and inter-basin water transfers a sustainable policy intervention for addressing water scarcity? *Water Security*, 9:100058.

Reznik, A., A. Dinar, S. Bresney, L. Forni, B. Joyce, S. Wallander, D. Bigelow, and I. Kan, 2021. Can managed aquifer recharge mitigate drought impacts on California's irrigated agriculture? The role for institutions and policies." *ARE Update*, 24(4):5–8.

Reznik, A., and A. Dinar, 2020. Reuse of recycled municipal wastewater by irrigated agriculture in the Escondido region, California. *ARE Update*, 23(4): 5–8.

Schaap, W. and F. van Steenbergen, 2001. *Ideas for Water Awareness Campaigns*. Stockholm, Sweden: The Global Water Partnership.

Suárez-Varela, M. and A. Dinar, 2020. The role of curtailment versus efficiency on spillovers among pro-environmental behaviors: Evidence from two towns in Granada, Spain. *Sustainability*, 12(3):769.

5 Environment–water interactions and management

The hydrological cycle that we have discussed so far includes the main water supply sources (e.g., precipitation) and the secondary derived sources (e.g., surface water in the forms of rivers and lakes, groundwater, recycled wastewater, and manufactured – desalinated – water). These sources are used by several sectors, such as irrigated agriculture to produce food and market products, the industrial sector to produce other market goods and households to produce benefits from their consumption (consumer surplus). In addition, water serves as an important input to sustain healthy water-dependent ecosystems. Water-dependent ecosystems provide valuable services to society, such as recreation and environmental quality (water and air quality).

Recent estimates of the value of ecosystem services highlight the importance of such services in economic terms. Services include provisioning services (e.g., food, water), regulating services (e.g., water quality, air quality, climate), cultural services (e.g., recreation) and habitat. de Groot et al. (2012) estimated ecosystem services values in terms of international dollars[1] per hectare of land per year in 2007 values. Using different groups of ecosystems, they suggest for water-related ecosystems an average value of $193,845 per average hectare of coastal wetlands, $25,682 per average hectare of inland wetlands and $4,267 per average hectare of rivers and lakes.

Watson et al. (2020) estimated that global ecosystem service values at 1.3 trillion international dollars in 2005 values. These values are subject to the effects of climate change and vary with the climate change scenario evaluated by the authors. Of these global values, water-related ecosystem services were estimated at $743,732 billion (57 per cent).

While the estimated values in the cited studies are hard to compare, both suggest that water-related ecosystem services are of significant high value to society. Therefore, the interactions between water and the environment are considered a high priority by regulators. One of the environmental policies used to regulate environment-water interactions is to secure scientifically based water flows. This chapter focuses on various ways that such interactions between water and the environment are being managed and regulated by environmental flow policies.

5.1   The concept of 'environmental flows'

A policy to secure the needed water resources for water-dependent eco-systems identifies a certain share of the flow to remain in the water system (river, lake) and is called *environmental flows, ecological reserve* or *water for the environment.* It is the share of total water resources in each system that sustains the water-dependent ecosystem and the ecological processes that guarantee the health of that system (Harwood et al., 2018).

For surface water systems, water for the environment is not simply a matter of quantity; it is a regime defined by 'the quantity, timing and quality of water required to sustain freshwater and estuarine ecosystems and the human livelihoods and well-being that depend on these ecosystems' (Brisbane Declaration, 2007).

In groundwater systems, for example, water for the environment is generally less well understood. However, it is recognized that the health of groundwater systems depends not just on water quantity but also on the timing of supply, quality and location of water (Murray et al., 2003). Consideration should also be given to conjunctive surface water–groundwater management in certain systems. More on environmental flows experiences can be found in Tickner et al. (2020).

5.2   How water affects the environment

When surface and groundwater systems experience alterations to natural water movement, distribution, temperature or quality, assets and ecosystems are impacted. Sometimes, those assets and ecosystems are pushed beyond their ability to recover, resulting in an irreversible process in the form of the degradation of the system as a whole. Changes to river flow regimes, for example, affect downstream uses, by changing natural systems along the river. Large or poorly timed abstractions from a single system can lower groundwater tables, lead to the drying of river mouths, destroy wetland habitats, prevent fish spawning and migration, reduce carbon cycling, increase pollution loads and lead to saltwater intrusion (Hirji and Davis, 2009).

While water for the environment does not explicitly include uses beyond ecological ones, it is recognized that the water that sustains ecosystems and associated ecological processes underpin the health of surface and groundwater systems. The health of those systems, in turn, sustains a wide range of social and economic benefits and values. By protecting or restoring water for the environment, various water-dependent social, cultural and economic values can be sustained. Conversely, unchecked use of water will impact the system health and all of the values that depend on a healthy system.

5.3   Management of water for the environment

Water for the environment can be secured either by external allocation of such water or by internal allocation, based on the ecological value function of the water. External allocation is based on external parameters that have been set and adjusted by scientists to a given water body or a stretch of that water body – usually a river (e.g., CHE, 2019). This approach relies on previous work by ecologists who estimated the sensitivity of different ecosystems in terms of level of services provided to different levels of water quantities and qualities during the year. Such prior information is used in reaching standards for environmental water flow levels that would be imposed on users in the different reaches of the river. Internal alloca-tion uses (when data is available) an estimated function of the response of the ecosystem to different levels of water availability (Esteban and Dinar, 2016), which is usually called the *ecosystem health function*. Inclusion of the ecosystem health function in addition to the other sectoral demands for water (e.g., agriculture and residential) in the river flow allocation optimization provides a comprehensive allocation for the entire river basin that takes into account the ecosystem demands for water and the resulting services.

When water for the environment is effectively planned and delivered with flexibility to adjust to changing physical conditions (such as availability of water and climatic conditions), it can contribute to realization of the wide range of environmental, social and economic benefits that can be derived from the use of water. It's increasingly clear that, in the mid- to long term, failure to meet environmental flow requirements has disastrous consequences for many water users (Dyson et al., 2003).

5.4 Examples of water allocations to address environmental flows

Several studies apply various approaches to analyze the value of environmental flows in different river basins around the world (e.g., Crespo et al., 2020; Dinar et al., 2008; Esteban and Dinar, 2016; Kahil et al., 2016). In the following I describe in more detail only two of these studies.

5.4.1 Allocation of environmental flows in the Ebro Basin, Spain

The Ebro Basin in Spain is characterized by having a diversity of uses along the very large area it spans, including irrigation projects in the northern parts of the basin, the very unique fish species in middle reaches of the river and special ecosystem needs at the bottom of the river where it flows into the ocean, including the Ebro Delta and fauna that is of high ecological value.

Crespo et al. (2020) developed a hydro-economic model for the Ebro Basin in order to evaluate water management policies that consider the interrelationships between economic activities and environmental uses. The hydro-economic model analyzes several sectoral and spatial water management policies under different water availability situations in the basin. The results show that policies that recognize the benefit of the environment achieve greater social well-being.

The hydro-economic model of the Ebro Basin includes hydrological, economic, environmental and institutional components in an integrated way. The model is composed of a hydrological model in a reduced form, a regional economy component and an environmental benefit component. Environmental flows are imposed on the different reaches of the river based on external scientific considerations.

The hydrological component represents the flows between the supply and demand nodes, using hydrological principles of mass balance and flow continuity in the river. It shows the spatial distribution of water flows used by the sectors and sub-regions. The regional economic component is composed of an optimization model of agricultural activities and of urban use. The agricultural activities model has been developed for each irrigation region and maximizes the private benefit of farmers from the production of crops subject to technical and resource constraints. The environmental component is characterized by the goods and services provided by the water-related environment, once appropriate water flows are available.

The Ebro River Basin Authority estimated the environmental flows needed to obtain different levels of ecosystem viabilities in various reaches of the basin. These were used as constraints to the model.

The model is used to analyze the economic and environmental effects of different drought management strategies. By varying the levels of the environmental flows (taking into account scenarios of drought and water availability) one can evaluate the intersectoral (irrigated agriculture, residential and environmental) outcome at the basin level. The three policy scenarios that are considered are:

1. Institutional cooperation: In drought conditions the basin authority reduces water allocations for irrigation proportional to the intensity of the drought. This allocation allows the distribution of water scarcity between agents and the environment. It's politics, which is currently applied in the Ebro Basin.
2. Water markets: The exchange of water is established between the irrigation communities, which means maximizing the private benefits of water use.
3. Environmental water markets: Under this policy, the basin authority has as its objective to maximize social welfare by maintaining the endowments that farmers receive under the institutional cooperation

policy. The basin authority participates in the market for water by acquiring the water necessary for the protection of aquatic ecosystems. The authority acquires water for ecosystems in an amount that maximizes social welfare, the sum of private benefits and environmental benefits. The environmental flow at the mouth is set at 3,000 hm^3 and stays on in all scenarios. The supply for urban use is guaranteed in all scenarios and it has priority over other uses, including the environment.

5.4.2   Environmental reserve flows in the Kat River, South Africa

The Kat Basin in South Africa demonstrates the need for a coordinated effort on the part of water users in order to sustain the different ecosystems along the Kat River, as well as the ecosystems beyond the Kat River. Dinar et al. (2008) demonstrate, using two approaches – negotiated role-playing and cooperative game theory – how the flow in the river can sustain three communities and ecosystems beyond the river. The focus in this chapter will only be on the game theory approach.

Most if not all water in the Kat Basin is from rainfall that is stored behind a dam and allocated among consuming sectors in villages and a town, more or less for the following uses: irrigation of citrus and vegetables, domestic consumption and ecological reserved flow for in-basin environmental needs and needs downstream to the main stem of the river. Given the stochastic nature of precipitation in the basin, some years pose challenges to the water availability for all uses. A basin water user association (WUA) regulates the annual allocations of water. The Kat Basin is divided into three sub-regions: upper (U), middle (M) and lower (L). It is assumed that the WUA a priori allocations are respected by all users – in other words, no water grabbing. Cooperation among the various users means that all respect the minimum ecological flow.

The model that was built can be described qualitatively as an objective function under constraints (Dinar et al., 2008: 99):

$$
\begin{array}{l}
\textit{Maximum water from the dam} \\
\textit{Water fee} \\
\textit{Water from the river}
\end{array}
\left\{
\begin{array}{l}
(\textit{land under citrus}); (\textit{land, water intensity cycles of cabbage}); \\
(\textit{water consumption in villages by source}); \\
(\textit{environmental releases})
\end{array}
\right\}
$$

Subject to:

1. environmental reserve outflow,
2. rainfall,
3. water flow in the river,
4. available land,
5. labor force, and
6. other institutional and technical constraints.

This optimization model was run for different combinations of the sub-regions, suggesting different levels of cooperation: the non-cooperation setup ({U}, {M}, {L}), where each sub-region maximizes their own benefits; the partial coalitions ({U, M}, {M, L}), where {U, L} is not operational due to the long distance between these two sub-regions, and the value created for U and L jointly is obtained by a linear summation of the value created by {U} plus the value created by {L}; and finally, the grand coalition of all regions ({U, M, L}). Because there is a dependency between all sub-regions, we can expect that empirically the best basin-level welfare (including the value from the environmental reserved flows) will be obtained when all sub-regions participate in a basin-wide arrangement that takes into consideration production, consumption and environmental needs. Numerical results can be found in Dinar et al. (2008).

Note

1. An international dollar would buy in a given country a comparable amount of goods and services that a US dollar would buy in the United States. This term can be seen as equivalent to a purchasing power parity (PPP) measure.

References

Brisbane Declaration, 2007. Environmental flows are essential for freshwater ecosystem health and human well-being. 10th International River Symposium and International Environmental Flows Conference (Brisbane, QLD). https://www.conservationgateway.org/ConservationPractices/Freshwater/ EnvironmentalFlows/MethodsandTools/ELOHA/Pages/Brisbane-Declaration .aspx.

Confederación Hidrográfica del Ebro (CHE), 2019. *Informe de seguimiento 2018-2019 del Plan Hidrológico de la Demarcación Hidrográfica del Ebro, ciclo de planificación hidrológica 2015-2021.* MITECO, Zaragoza.

Crespo, D., J. Albiac, A. Dinar, E. Esteban and T. Kahil, 2020. Beneficios medioambientales de los ecosistemas en el modelo hidroeconómico de la Cuenca del Ebro. Unidad de Economía Agroalimentaria Centro de Investigación y Tecnología Agroalimentaria, Documento de Trabajo 20/02. https://citarea.cita-aragon.es/citarea/bitstream/10532/5108/1/2020_230.pdf.

de Groot, R., L. Brander, S. van der Ploeg, R. Costanza, F. Bernard, L. Braat, M. Christie, N. Crossman, A. Ghermandi, L. Hein, S. Hussain, P. Kumar, A. McVittie, R. Portela, L. C. Rodriguez, P. ten Brink, and P. van Beukering, 2012. Global estimates of the value of ecosystems and their services in monetary units. *Ecosystem Services*, 1:50–61.

Dinar, A., S. Farolfi, F. Patrone and K. Rowntree, 2008. To negotiate or to game theorize: Evaluating water allocation mechanisms in the Kat Basin, South Africa. In A. Dinar, J. Albiac, and J. Sanchez-Soriano (eds.), *Game Theory and Policy Making in Natural Resources and the Environment*, ch. 5. London: Routledge Publishing.

Dyson, M., Bergkamp, G. and Scanlon, J. (eds.), 2003. *Flow: The Essential of Environmental Flows.* IUCN, Gland, Switzerland and Cambridge, UK.

Esteban, E. and A. Dinar, 2016. The role of groundwater-dependent ecosystems in groundwater management. *Natural Resource Modeling*, 29(1):98–129.

Harwood, A. J., D. Tickner, B. D. Richter, A. Locke, S. Johnson, and X. Yu, 2018. Critical factors for water policy to enable effective environmental flow implementation. *Frontiers in Environmental Science*, https://doi.org/10.3389/fenvs.2018.00037.

Hirji, R. and R. Davis, 2009. *Environmental Flows in Water Resources Policies, Plans, and Projects: Case Studies.* Environment Department Papers, #117. Washington DC: World Bank.

Kahil, M. T., A. Dinar, and J. Albiac, 2016. Cooperative water management and ecosystem protection under scarcity and drought in arid and semiarid regions. *Water Resources & Economics*, 13:60–74.

Murray, B. R., M. J. B. Zeppel, G. C. Hose, and D. Eamus, 2003. Groundwater-dependent ecosystems in Australia: It's more than just water for rivers. *Ecological Management & Restoration*, 4(2):110–13.

Tickner, D., N. Kaushal, R. A. Speed, and R. E. Tharme (eds.), 2020. *Implementing Environmental Flows: Lessons for Policy and Practice.* Lausanne: Frontiers Media SA.

Watson, L., M. W. Straatsma, N. Wanders, J. A. Verstegen, S. M. de Jong, and D. Karssenberg, 2020. Global ecosystem service values in climate class transitions. *Environmental Research Letters*, 15:024008.

6 Economic and policy considerations in groundwater management

Groundwater (GW) is one of the main water resources on our planet. We already discussed the distribution of water sources on Earth in Chapter 2. Groundwater is defined as water stored in aquifers beneath the surface (Shaw, 2021). Two main distinct types of groundwater include replenished aquifers and fossil aquifers. Replenished aquifers can be recharged by precipitation, return flows from irrigated agriculture and artificial recharge. A more detailed explanation on the physics of groundwater can be found in Shaw (2021: 206–10).

In recent years, GW issues have surfaced and involve many local and international agencies that attempt to address the problems faced by nations in managing aquifers storage and uses (United Nations WATER, 2018). With the impact of climate change on the sustainability of GW storage, scientists realized that GW level changes correspond to selected global climate variations (Gurdak, 2017; Russo and Lall, 2017; Thomas and Famiglietti, 2019).

An assessment of the National Ground Water Association (NGWA, n.d.) suggests that 0.61 per cent of all water on Earth (which is 94 per cent of available freshwater, after deducting ocean water and glaciers and other ice forms) is extracted from aquifers, confirming the significant role of GW in our planet.

But the extent of GW usage across countries varies. Table 6.1 presents data from major GW-using countries.

The data in Table 6.1 suggests that the availability of GW differs in different countries, stemming probably from the depth and quality of the water. Another result in Table 6.1 is that irrigated agriculture is the major consumer of GW in the world, which is reflected also in the negative effects of pollution from agricultural activities, and their regulation, as we will discuss later in the chapter.

The massive extraction seen in Table 6.1 is not typical only to the seven countries listed in the table. Extraction of groundwater has increased exponentially over time almost everywhere in the world.

Scrutiny of the global GW extraction curve in Figure 6.1 suggests that, in the past century, extraction of GW has increased nearly 4,000-fold. This of course cannot go without short- and long-term consequences. In the next sections we will review the implications of GW extractions and various economic and policy responses.

Given the significant role of GW in buffering the environment and socio-economic activities against stochastic and permanent water scarcity and drought, it is important to realize the various negative impacts of unsustainable management of GW and policies to address such impacts (Kath and Dyer, 2017).

Table 6.1   Largest annual groundwater extracting countries and sectoral distribution (2010)

| Country | Groundwater extraction (estimates) | | | |
| | GW extraction (km^3/year) | Breakdown by sector (%) | | |
		Irrigation	Domestic	Industry
India	251.00	89	9	2
China	111.95	54	20	26
United States	111.70	71	23	6
Pakistan	64.82	94	6	0
Iran	63.40	87	11	2
Bangladesh	30.21	86	13	1
Mexico	29.45	72	22	6

Source: Adapted from Margat and van der Gun (2013).

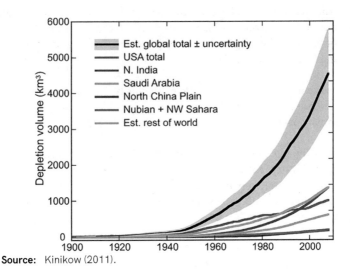

Source: Kinikow (2011).

Figure 6.1 Global groundwater extraction over time (1900–2010)

6.1   Implications of groundwater extraction on users and the environment

It is a well-known physics law that when water is taken out of a container (aquifer), without any replenishment, the level of the water in the aquifer will draw down. This drawdown may lead to various implications, depending on many parameters, including the characteristics of the aquifer and its location. Aquifers around the world, especially in semi-arid regions where natural replenishment has been irregular, face increased levels of extractions as climate change scarcity and drought periods exceed in occurrence. After all, when surface water supplies are curtailed during droughts, water users turn to the 'water-saving bank' for help. This was exactly the process that led the state of California, following a sequence of drought years, to face depleted GW aquifers and introduce the 2014 Sustainable Groundwater Management Act (SGMA), where new legislation was introduced to change the institutional setting of GW management in the state, which is expected to be a more responsible management of this resource and allow for coping with drought years (Water Education Foundation, n.d.).

6.1.1   Depth to water table

As water is pumped out the aquifer, the level of the water table draws down until water in the aquifer is replenished. This has an immediate effect on the cost of pumping for the users who face a deeper level from which they need to pump the resource. The economic effect on profitability of agricultural production is amplified in the case of poor farmers who do not have resources to deepen their wells and, thus, become dependent on wealthy farmers who pump and distribute the water to them (Feinerman and Knapp, 1983).

In addition to affecting the energy cost of pumping, lower water table in a given aquifer may lead in the long run to flow of water into that aquifer from surrounding aquifers, in many cases with lower water quality (mainly with high salinity content). When such affected aquifers are located next to the ocean, lowering of their water table may lead to sea-water intrusion (Darcy's law), which results in water quality deterioration. This phenomenon also has economic consequences, where water may no longer be adequate for its intended use, or use of such water may lead to damages in the conveyance system (from salinity) and yield losses if the water was used for irrigation (Zeitouni and Dinar, 1997).

6.1.2   Land subsidence

Land subsidence (LS) is the phenomena of a sudden or gradual sinking of the land surface in response to either natural processes, such as oxidation and vaporization of organic soils, or human activities such as oil mining or groundwater extraction, among others. Subsidence is a global problem, known to be associated with long-term groundwater over-pumping of unconsolidated alluvial aquifers composed of a substantial fraction of fine-grained sediments.

It is assessed that LS inflicts significant damages on local communities and on the environment, such as infrastructure damage and collapse, soil fracture leading to loss in life in some cases, reduced performance of hydrological systems such as loss of storage capacity, and malfunctioning of drainage systems, to name a few (Dinar et al., 2021). Management of GW with consideration of both direct LS damages and indirect damages, such as loss of storage capacity, would reduce and prevent LS's negative effects.

6.1.3   Greenhouse gas emissions from depleted groundwater aquifers

As suggested by Wood and Hyndman (2017), a significant and previously unrecognized source contributing to CO_2 emission is from groundwater depletion, which is released when water is removed from soil and rocks. For example, it is estimated that an average 1.7 million metric tons of CO_2 is released in the United States from depleted groundwater annually. Preventing water table drop below certain levels may reduce CO_2 emission from aquifers. In that respect, GW depletion can be seen as a result of climate change, but also as affecting climate change.

6.2   Groundwater quality deterioration and impact on human health

In addition to negative externalities in the form of water pumping congestion and land subsidence, GW is associated with other externalities impacting human health that require government intervention (Lall et al., 2020). Those negative health impacts include nitrates and perfluoroalkyl and polyfluoroalkyl substances (PFAS) contaminations. Regulation of agricultural-source contamination of GW has been one of the most challenging policy interventions in many countries around the world.

6.2.1   Nitrates

Groundwater is a major source of nitrate contamination in drinking water resulting from the use of nitrate-based fertilizers and other chemical during the agricultural production process. Ward et al. (2018) found evidence of a relationship between drinking water nitrate ingestion and adverse health outcomes, including adverse reproductive outcomes and other health effects mainly in the form of colorectal cancer, thyroid disease and neural tube defects. Preventing excessive use of fertilizers in agricultural production may help regulate leaching of nitrates to groundwater.

6.2.2   PFAS

Reclaimed water, residential treated wastewater, is becoming an increasingly important source of water for irrigation in arid regions worldwide.

Wastewater is used to recharge the aquifers and irrigate parks, golf courses, recreational fields and certain agricultural crops. Concentrations of perfluoroalkyl and polyfluoroalkyl substances (PFAS) are high and may affect human health. PFAS are contaminants of emerging concern because of their widespread occurrence, toxicological impact to humans and persistence in the environment (Cáñez et al., 2021). Regulation of recycled wastewater use in irrigated agriculture, or production of higher levels of treatment (leading to more expensive recycled water), can be considered as a solution to PFAS in groundwater.

6.3   Policy interventions to address implications from groundwater over-extraction

The direct and indirect negative externality effects associated with groundwater depletion and contamination on the environment and on humans call for regulatory interventions. Groundwater is an open-access resource (allowing anyone above the aquifer to use its water), and thus it introduces management complications that stem from the nature of water extraction under common-pool behavior of the users. One possible management of groundwater is a joint management by all users, voluntarily internalizing the negative externalities and maximizing all users' welfare. In addition, external regulators may address over-pumping or pollution and their consequences by various interventions, including standards, taxes, incentives and institutions. Most policy interventions aim to regulate groundwater extractions and agricultural production above an aquifer.

6.3.1   Caps on water extractions

Caps (constraints) on water pumping is a common approach that requires information on equipment capacity and hours of operation. Knowing the safe yield of the aquifer (the amount of water that can be used without depleting the level of water in the aquifer over time) is a necessary requirement for setting the cap on extraction. Then the total amount of the cap is distributed in one way or another among the users, also allowing trade in the quantities of water caps (Madani and Dinar, 2012).[1]

Information is not always available to the regulator, especially on individual extractions from the aquifer. Under asymmetric information the regulator may consider either investment in information gathering (Dinar and Xepapdeas, 2002), for example by estimating a relationship between the type of the irrigated crops and the size of the pump used, in order to estimate the amount of water extracted (Dinar, 1994). The regulator may also employ advanced methods of data collection on groundwater use, such as employing periodical (even twice daily) satellite images to assess groundwater use on farms (Bastiaanssen and Hellegers, 2007). When information on extraction above the cap becomes known to the regulator, it allows the introduction of consequences such as payments of fines, or reduction of the level of the cap in the following period.

The case of the Sustainable Groundwater Management Act, California

Groundwater in California has not been regulated formally by the state. Many groundwater users could, for many years, dig wells into their land and pump as much water as they needed. The long drought faced by California since 2010 has led many of the main aquifers in the state to draw down with all the consequences that we discussed above.

The historic passage of the SGMA in 2014 introduced a statewide legal and operational framework that helps protect groundwater resources in a sustainable manner. Without getting into the details of SGMA, its principles emphasize that groundwater management in California is to be accomplished locally. After identifying GW basins of low, medium and high priority in terms of their water situation, local agencies are requested to form groundwater sustainability agencies (GSAs) for the high- and medium-priority basins. These GSAs develop and implement groundwater sustainability plans (GSPs) to prevent undesirable water depletion results and mitigate their overdraft within 20 years. The principles by which SGMA operates are similar to those we discussed in this section (AquaOSO, n.d.). In addition, SGMA promotes joint management of aquifers (Section 6.3.5) and institutional reforms for the management of aquifers (Section 6.3.6).

6.3.2   Taxes on water extractions

Other means of intervention and regulation of water extraction are Pigouvian taxes (see Chapter 7 for more explanation on use of taxes for pollution regulation) imposed on incremental extraction by equating the marginal damage from extraction to the tax level. Effective Pigouvian taxes can be imposed on users only if volumetric measurement of the extraction is in place. Extraction taxes can be uniform (where the tax per quantity extracted is the same for all levels of extraction) or in increased blocks (where the tax per unit of extraction increases with the level of extraction). The idea behind such tax is to internalize the scarcity value of the extracted water by the users and affect their extraction decisions (Esteban and Dinar, 2013a).

6.3.3   Taxes on energy used for pumping

A different policy that takes advantage of metered energy used for pumping groundwater is to impose a tax on the energy (usually electricity) used for pumping. In this case the regulator imposes a tax (either uniform or increased blocks) on the electricity as measured in the meter of the user.

In a study in Northern Gujarat, India, Fishman et al. (2016) found that farmers' participation in a program with charges on electricity use for pumping has shown no impact on levels of water extraction. In a different study, Tellez-Foster et al. (2017a, b) found that moving from subsidized electricity rates for irrigation water pumping to a less subsidized or non-subsidized electricity rate led to reduced extraction rates in an over-drafted aquifer in Mexico. While the objective of the study in Mexico was to find how groundwater users would respond to a reduction or elimination of electricity subsidy, still the study found that such policies affect the groundwater extraction decisions of the water users and thus tell us about the effectiveness of energy taxes.

6.3.4   Investment in information and subsidies to use efficient (irrigation) technologies

For quite some time the notion of efficient irrigation technologies has been related to a water conservation effect. However, several studies (Ward and Pulido-Velazquez, 2008; Fishman et al., 2015) have shown that the expansion effect, where water saved by use of more efficient technologies, culminates in expansion of irrigated land and not returning the saved water to the aquifer. Fishman et al. (2015) empirically find that widespread adoption of improved irrigation technologies (drip and sprinkler irrigation) has the potential to reduce the amount of excessive extraction of groundwater by two-thirds. However, under more realistic assumptions about farmers' irrigation choices, half of these reductions are lost due to the expansion of irrigated area (when land is not constrained). Therefore, without the introduction of incentives for conservation, much of the potential impact of efficient technology adoption on aquifers may be lost.

Dinar and Xepapadeas (1998, 2002) studied an aquifer in Karen County, California. Given the nonpoint source pollution nature of groundwater, two policy interventions have been considered in order to expand its informational base: a tax scheme and an investment in information on farmer extraction from and pollution of the aquifer. Both are designed to obtain an optimal level of aquifer water use and emissions to the aquifer by a group of agricultural producers whose individual water use and emissions from and to the aquifer cannot be known without investing in monitoring. The regulatory agency invests in increased observability of water extraction and salinity emissions, and then, once information is available, taxes different levels of extractions. It is obvious that investment in increasing data availability should be equal to the prevention of pollution damages plus damages from over-extraction.

6.3.5 Cooperative management practices

So far, we have discussed top-down policies that are imposed on the users by a regulatory agency, if the individual users respond to the price or quota signals sent by the regulatory agency in such a way that considers their own situation. However, users of an aquifer can organize to respond cooperatively, so that they maximize the long-term benefits of the entire group of users. This approach implements components of cooperative game theory. It can operate under several assumptions about the interactions between the users (agents). Acting cooperatively in the case of an open access resource may be preferred to a competitive approach in managing such resource.

Esteban and Dinar (2013b) analyzed cooperative policy interventions in the Eastern La Mancha, Spain, imposed and managed by the farmers' Water Users Association. The analysis considered cooperation among three sub-regions across the surface area of the aquifer who may be coordinating their agricultural production operations. Different cooperation arrangements were considered: (1) the status quo, in which farmers from each sub-aquifer maximize their private gross margin without internalizing the extraction externality and each sub-aquifer is modeled separately; (2) partial cooperation, in which farmers from neighboring sub-aquifers cooperate in groups of two (with the closer neighbor), internalizing the extraction externalities (the farmers from the third sub-aquifer, not in the coalition, act as 'free-riders' without internalizing the water extraction externality); and (3) full cooperation in which a coalition between the farmers of the three sub-aquifers is simulated with the extraction externality being internalized by farmers from all regions. The results of the simulations show, as expected, that individual farmers and the entire society are always better off with a cooperative rather than when than when farmers act individually. Furthermore, when an environmental externality (such as loss of ability of the water-related ecosystem to produce benefits) is considered in the GW model, the cooperative results are more appealing than the non-cooperative results.

6.3.6   Packaging and sequencing various policy interventions

There is evidence that a policy composed of a set of intervention measures may lead to better results than a policy based only on one intervention measure. Esteban and Dinar (2013a) applied the concept of packaging and sequencing policy interventions aimed at regulating a given aquifer. They hypothesized that a policy composed of a package of quota and tax interventions that are also sequenced in the timing of their implementation would result in a higher regional welfare than a quota policy only or a tax policy only.

They applied a policy that combines both a quota on extraction and a uniform water tax imposed on farmers that use the Eastern La Mancha aquifer water. In a dynamic recursive model, they allowed the regulator to compare the net present value of regional welfare calculated over a planning period and resulting from a combined policy (quota and tax), an individual quota and tax policies, and a sequencing structure of the quota and tax interventions (tax and quota, or quota and tax introduced one after the other). Both the quota and the water tax levels in the packaged policy are similar to the levels simulated in the separate applications of tax and quota.

The implementation of this package of policy instruments shows interesting patterns. Without getting into specific results, it is evident that a selection of tax or quota components of the policy are a function of the relative effect of the physical outcomes of the aquifer water levels. What is seen is a switching between quota and tax, depending on their relative effectiveness during the planning horizon. In comparison to the (social planner)[2] efficient policy intervention, the packaged and sequenced quota and tax instruments are considered (in Esteban and Dinar, 2013a) a second best, achieving 98.8 per cent of the social planner value of the regional social welfare. The uniform tax alone is a third best, with 98.5 per cent of the social planner value.

6.4   Discussion

Management of GW resources has challenged policymakers for many years. The physical characteristics that make GW a hidden and difficult-to-monitor resource, as well as being an open access common-pool resource, lead to challenges in managing it effectively. Indeed, as we have seen, GW stocks around the world have been reduced over time, also experiencing a reduction in their quality.

Despite these challenges, management of GW aquifers is undertaken by regulatory agencies, with different levels of effectiveness. We distinguished between top-down regulations like imposing quotas on extraction and taxes on level of extraction (either directly on water pumped, or indirectly on electricity used to pump water). These are only part of the policy intervention mechanisms used. We also mentioned the possible trade in groundwater quotas among the users that have been awarded water rights from the aquifer. We also covered cooperative institutions that are established by users of the GW aquifer and mechanisms to sustain these cooperative institutions.

We have not compared all these policy interventions and ranked them in terms of their effectiveness. We leave this part to a higher level of analysis that could be performed at a different scale and extent (see for example Dinar and Tsur, 2021; Shaw, 2021).

Notes

1. The process of limiting the aquifer extraction and distributing quotas among users is practiced in California under the title *adjudication*. It is managed by the court and is typical in several large aquifers in Southern California.
2. A social planner solution is considered a *first-best*. It is obtained by a decision-maker who attempts to achieve the best result for all parties/users involved in the utilization of the aquifer water.

References

AquaOSO, n.d. SGMA California. Accessed 10 December, 2021. https:// aquaoso.com/resources/sgma-california/?utm_term=sgma&utm_campaign= California&utm_source=adwords&utm_medium=ppc&hsa_acc=3448002798 &hsa_cam=11003487528&hsa_grp=113658253181&hsa_ad=492863222319& hsa_src=g&hsa_tgt=kwd-339446574635&hsa_kw=sgma&hsa_mt=b&.

Bastiaanssen, W. G. M. and P. J. G. J. Hellegers, 2007. Satellite measurements to assess and charge for groundwater abstraction. In A. Dinar, S. Abdel Dayem, and J. Agwe (eds.), *The Role of Technology and Institutions in the Cost Recovery of Irrigation and Drainage Projects*. Discussion Paper 33. World Bank: Agriculture and Rural Development Department.

Cáñez, T. T., B. Guoa, J. C. McIntosh, and M. L. Brusseau, 2021. Perfluoroalkyl and polyfluoroalkyl substances (PFAS) in groundwater at a reclaimed water recharge facility. *Science of The Total Environment*, 791:147906.

Dinar, A., 1994. Impact of energy cost and water resource availability and quality on agriculture and groundwater quality in California. *Resources and Energy Economics*, 16:47–66.

Dinar, A. and Y. Tsur, 2021. *Economics of Water Resources: A Comprehensive Approach*, Cambridge, UK: Cambridge University Press.

Dinar, A. and A. Xepapadeas, 1998. Regulating water quantity and quality in irrigated agriculture. *Journal of Environmental Management*, 54:273–89.

Dinar, A. and A. Xepapadeas, 2002. Regulating water quantity and quality in irrigated agriculture: Learning by investing under asymmetric information. *Environmental Modeling and Assessment*, 7(1):17–27.

Dinar, A., E. Esteban, E. Calvo, G. Herrera, P. Teatini, R. Tomás, Y. Li, P. Ezquerro, and Jose Albiac, 2021. We lose ground: Global assessment of land subsidence impact extent. *Science of the Total Environment*, 786:147415.

Esteban, E. and A. Dinar, 2013a. Modeling sustainable groundwater management: Packaging and sequencing of policy interventions. *Journal of Environmental Management*, 119:93–102.

Esteban, E. and A. Dinar, 2013b. Cooperative management of groundwater resources in the presence of environmental externalities. *Environmental and Resource Economics*. 54:443–69.

Feinerman, E. and K. Knapp, 1983. Benefits from groundwater management: Magnitude, sensitivity, and distribution. *American Journal of Agricultural Economics*, 65(4):703–10.

Fishman R., N. Devineni, and S. Raman, 2015. Can improved agricultural water use efficiency save India's groundwater? *Environmental Research Letters*, 10:084022.

Fishman R., U. Lall, V. Modi, and N. Parekh, 2016. Can electricity pricing save India's groundwater? Field evidence from a novel policy mechanism in Gujarat. *Journal of the Association of Environmental and Resource Economists*, 3:819–55.

Gurdak, J. J., 2017. Groundwater: Climate-induced pumping. *Natural Geosciences* 10:71.

Kath J. and F. J. Dyer, 2017. Why groundwater matters: An introduction for policy-makers and managers. *Policy Studies*, 38:447–61.

Kinikow, L. F., 2011. Contribution of global groundwater depletion since 1900 to sea-level rise. *Geophysical Research Letters*, 38(17). https://doi.org/10.1029/2011GL048604.

Lall, U., L. Josset, and T. Russo, 2020. A snapshot of the world's groundwater challenges. *Annual Review of Environment and Resources*, 45:171–94.

Madani, K. and A. Dinar, 2012. Cooperative institutions for sustainable common pool resource management: Application to groundwater. *Water Resources Research*, 48(9). https://doi.org/10.1029/2011WR010849.

Margat, J., and J. van der Gun, 2013. *Groundwater around the World*. London: CRC Press/Balkema.

NGWA (National Ground Water Association), n.d. Information on earth water. Accessed 18 November, 2021. https://www.ngwa.org/what-is-groundwater/About-groundwater/information-on-earths-water.

Russo, T. A. and U. Lall, 2017. Depletion and response of deep groundwater to climate-induced pumping variability. *Natural Geosciences* 10:105–108.

Shaw, W. D., 2021. *Water Resource Economics and Policy: An Introduction*, 2nd ed. Cheltenham, UK and Northampton, MA, USA: Edward Elgar Publishing.

Thomas B. F. and J. S. Famiglietti, 2019. Identifying climate-induced groundwater depletion in GRACE observations. *Scientific Reports*, 9:4124.

Tellez-Foster E., A. Rapoport, and A. Dinar, 2017a. Groundwater and electricity consumption under alternative subsidies: Evidence from laboratory experiments. *Journal of Behavioral and Experimental Economics*, 68:41–52.

Tellez-Foster E., A. Rapoport, and A. Dinar, 2017b. Alternative policies to manage electricity subsidies for groundwater extraction: A field study in Mexico, *Journal of Behavioral Economics for Policy*, 2(2):55-69.

United Nations WATER, 2018. *Groundwater Overview: Making the Invisible Visible*. Delft, Netherlands: International Groundwater Resources Assessment Center.

Ward M. H., R. R. Jones, J. D. Brender, T. M. De Kok, P.J. Weyer, et al., 2018. Drinking water nitrate and human health: an updated review. *International Journal of Environmental Research and Public Health*, 15:1557.

Ward, F. A. and M. Pulido-Velazquez, 2008. Water conservation in irrigation can increase water use. *Proceedings of the National Academy of Science, U.S.A.*, 105:18215.

Water Education Foundation, n.d. Aquapedia background: Sustainable Groundwater Management Act (SGMA). Accessed 26 February, 2022. https://www.watereducation.org/aquapedia-background/sustainable-groundwater-management-act-sgma.

Wood, W. W. and D. W. Hyndman, 2017. Groundwater depletion: A significant unreported source of atmospheric carbon dioxide. *Earth's Future,* 5:1133–35.

Zeitouni, N. and A. Dinar, 1997. Mitigating negative water quantity and quality externalities by joint management of adjacent aquifers. *Environmental and Resource Economics,* 9:1–20.

7   Economics of water pollution regulation

Water is an essential input in the production of many goods (and services). However, the production process of many, if not all, goods also include by-products – pollutants. In many cases, these pollutants infiltrate water bodies, either surface or underground, and may lead to severe consequences in the form of deteriorated quality of these waters. High levels of pollution of water resources may make them not useable for their original purpose, or even for any purpose. In extreme cases, high pollution levels of water may lead directly or indirectly to health issues (see Chapter 6). Water pollution is a negative externality because, even if it is not intended by the polluters, the effects on other water users are significant and lead to social welfare loss. As such, governments intervene in order to regulate pollution of water resources. In this chapter we review and analyze the various policies used by regulatory agencies to address various types of water pollution. Readers interested in more details on global pollution and its consequences are referred to Mateo-Sagasta, et al. (2017) for pollution from the agricultural sector and to Schwarzenbach et al. (2010) for global health impacts of water pollution.

There are many sectoral sources for pollution and many effects of water pollution on humans and the environment. A recent study (UNEP, 2016) identified types of pollution sources and their impacts. Pollution of water is not typical only to developing countries, where water quality regulations are loose or non-existent. Depending on the type of water pollution (Pacific Institute, 2010), one can actually observe increase in water pollution as industrial activity becomes more intensive. Generally speaking, water pollution can be the result of household water use, agricultural water use and industrial and mining water use. I will restrict the discussion in this chapter to water pollution by the residential sector and the agricultural sector.

An important source of water pollution is residential human wastes that in many countries are dumped into the water system and affect human and ecosystem health. Other important sources of water pollution are residuals from the agricultural production process in the form of salinity or chemicals resulting from application of pesticides and fertilizers, and in many cases, also the result of natural processes embedded in the local soil. Finally, there is water pollution from the industrial and mining sectors dumping highly polluted chemicals from different industrial processes. While each type of pollution has its own unique characteristics and effects on human and ecosystem health, the principles related to their regulation could be quite similar. Below I provide general principles for water pollution regulation with examples taken from these three sectors.

7.1   Principles for regulating water pollution

Several features are considered by the regulator when developing and implementing a pollution-reduction intervention in a given water body. The regulator opts to minimize the social cost associated with pollution of that water body. What are the components that need to be considered in each policy intervention? The damage from pollution to human health and to ecosystems is one component; the abatement cost associated with the reduction of the pollution of the water body is another component; and the transaction costs of monitoring and enforcing the regulation is a third component. By taking all these components into consideration, the policy maker reaches an optimal pollution/abatement level that is best for the society that uses the water body in question.[1] The next section describes the various policy mechanisms that could be employed in order to reduce the level of water pollution.

7.2   Mechanisms for regulation of water pollution

While we refer to regulatory mechanisms in the context of water pollution, such mechanisms can be employed for any type of pollution (e.g., air, noise, nitrates). We refer to five groups of policy mechanisms: (1) standards, (2) emission taxes, (3) indirect taxes (4) abatement subsidies, and (5) tradable discharge permits.

7.2.1   Standards

Setting water quality standards for compliance by producers/consumers is the traditional approach to regulating pollution, including water pollution. The basic principle is to establish standards for polluters that will achieve the desired level of pollution (in a particular location/society) and penalize polluters for not meeting such standards. This approach is often called 'command and control' because polluters are told what to do.

There are several types of standards. First, technology standards dictate to producers/users what kind of production/consumption or abatement technology should be used. For example, irrigators might be directed to use drip irrigation in a given location or on given crops; residential water users might be enforced to use certain types of toilets in their bathrooms. On the one hand, technology standards are relatively easy to monitor and enforce by the regulator, but on the other hand, they make it difficult to achieve cost-effective abatement because the regulator is not always familiar with all available technologies and might not have selected the most efficient one. Technology standards provide little incentive for technology innovation by the polluters since they would prefer to stick to the standard. Since each firm (or household) has its own production/consumption constraints, it would be hard to provide technology standards for each firm/household.

Second, emission standards allow the emission of a certain level of a given pollutant. Emission standards are dictated by a federal or local water quality agency (such as the federal Environmental Protection Agency (EPA) or the state EPA). It is questionable whether a blanket standard or a quilt standard policy performs better, but this question will not be discussed and addressed in our chapter. Emission standards require that a firm emits no more than a specified amount of a pollutant and leaves the decision of the technology (production or abatement) to the polluter. Firms/households make technology choices on their own, which is an incentive for cost-effective abatement via technology innovation.

Third, ambient standards require that the level of a pollutant be no more than a specified level in a given locality. Examples may include air quality level in a given region, or level of nitrates in a given GW aquifer. Given the level of the ambient standard, firms make technology choices, which provides incentives for cost-effective abatement. One economic/social question that comes to mind is whether ambient standard (quilt regulation) is more, or less, equitable than a national standard (blanket regulation)? A national standard for water quality may be more (transaction)-cost effective in terms of monitoring but may not be overall socially effective in terms of abatement cost. Clearly this is an important economic-policy consideration.

To summarize, standards are set to establish pollution limits and/or technology requirements and enforce them with penalties if/when they are violated. Benefits of standards include their simplicity in terms of being straightforward and easy to understand by polluters. They appeal to certain notions of fairness, and address the common feeling that pollution is undesirable and ought to be restricted directly by society. Drawbacks of standards (depending on the specific standard) include that it is not always possible to monitor them; that it is difficult to achieve cost-effective abatement in the short run (polluters are not able to adjust their technology fast enough); and that it is questionable whether standards provide incentives for abatement innovation in the long run.

7.2.2   Emission taxes

So far, we assumed that environmental resources (such as water) and services (amenities) tend to be freely available, which is a fundamental cause of their overuse and pollution. An emission standard, for example, is a quantity instrument in the sense that it limits the quantity of pollution directly. An alternative policy intervention instrument – a tax (price) instrument – requires users/polluters to pay for their use and abuse of environmental resources and services. Emission taxes are a relatively new approach to regulating environmental pollution compared with standards.

The idea behind emission taxes is to allow pollution sources to emit as much as they want to, but to charge them a fee per unit of emission. The economic principle behind such a mechanism is that polluters face an abatement cost function that assigns a cost per each incremental unit of the pollutant that is removed from being emitted (a marginal abatement cost, or MAC). The tax (fee) level imposed on the polluter can be designed by the regulatory agency in such a way that it will be equal to the MAC at the desired level of pollution. Each polluter facing a given level of the emission tax will decide on how much to pollute and how much to abate. The total sum associated with the emission tax equals the sum of the actual tax paid by the polluter, the abatement costs incurred by the polluter and the tax burden (which is a sum that the polluter pays as a tax even on pollution prevented).

To summarize, the benefits of emission tax can be seen in its cost effectiveness, due to the fact that the marginal abatement cost is equalized to the tax level (we call it the *equimarginal principle*). Emission tax also seems fair since polluters must pay to use resources or environmental services. Users pay as they pollute, which is a free choice. Emission tax provides strong incentives for abatement innovation. By innovating on abatement cost, the polluter reduces the slope of the marginal abatement cost function. In doing so, the tax burden is reduced more than the abatement cost.

An important policy question is what the regulatory agency does with the sum of the tax that was collected: Is it returned to the society (redistributed) to improve the way it handles pollution?

Emission taxes have several drawbacks. They can be objected to due to moral arguments opposing the commoditization of environmental quality. Under emission taxes, it is possible that 'hot spots' of pollution will be created, depending on the level of the tax that was imposed by the regulatory agency. Emission tax also creates higher costs for polluters compared with standards. Under emission taxes the polluter faces abatement costs plus the tax burden – under a tax, firms have to pay for the pollution they get rid of and for the pollution they still generate (not so with a standard). This is why firms may lobby for standards.

7.2.3   Indirect taxes to regulate nonpoint source pollution

What if it's politically or technically impossible to impose a tax on the emitted volume of polluted water? This may happen when the regulator has no way to assess the level of pollution emitted by a given polluter. This is the situation with agricultural pollution, where emission cannot be identified with given fields or farms. Should we instead tax something that is closely related to the process leading to the creation of the pollutant in question?

For example, in the case of nitrate in the drainage water, which is a hazardous pollutant, the regulatory agency could apply a direct policy of taxing the nitrogen (concentration) in the drainage water. However, since there is no way to monitor the exact level of nitrogen from a given field, an alternative (second best) emission tax can be introduced via the taxing of fertilizers containing nitrogen that are used as an input in the agricultural production process on that field.

7.2.4   Abatement subsidies

While emission taxes make polluters pay a fee for doing something that is socially undesirable (MAC as a tax), the principle of an abatement subsidy is quite the opposite, namely, pay polluters for doing something that is socially desirable (MAC as a subsidy) – invest in pollution-reducing technologies. The result of these two policy interventions is expected to be similar in terms of the individual polluter behavior. However, they are not entirely the same. The regulator should be concerned about possible macro-economic consequences on the industry. Emission tax reduces profits of the individual firms and leads firms to leave the industry. Therefore, both individual and total industry emissions will decrease.

The case of an emission subsidy is quite the opposite. Subsidy increases profits, making the subsidized industry attractive and leading firms to enter the industry. While individual emissions may decline due to the emission-reducing technology that was introduced, total industry emissions may increase due to the higher number of firms entering the industry due to the subsidy.

7.2.5   Tradeable discharge permits

A relatively new concept of water pollution regulation is tradable discharge permits (TDP), building on the principles of water trade. The regulatory agency in charge of the control of water quality in a given region or stretch of the river needs to follow certain steps. First, a reliable assessment of the level of discharge to the water body under regulation has to be obtained. Then, an assessment of the damage from pollution that has to be prevented and the desired level of pollution have to be estimated. The difference between the actual level of pollution and the desired level of pollution is called a *cap* and is made public. That cap has to be divided or allocated among all polluters by means of giving, drawing or purchasing. This creates a new property right – the right to emit certain number of units of pollution to the water body that is the property of the permit holder. The regulatory agency then needs to set a mechanism that allow polluters that hold permits to trade these rights with one another. It is expected that polluters with high abatement costs will buy permits from sources with lower abatement costs. A price per permit will emerge in the permit market and the desired level of pollution will be reached.

TDP provides an automatically cost-effective abatement mechanism and provides strong incentives to polluters to innovate by making their abatement process more efficient and thus improve their competitiveness in the market for permits. TDP has a low information requirement and the regulatory agency doesn't need to be involved in the process after setting the level of cap and assigning the property right levels for permits. The payment burden on polluters tends to be smaller than for an emission tax. Some of the disadvantages include the need for monitoring that the market performs well and that the level of pollution is kept below the desired level of pollution. With TDP there is still a possibility for having pollution hot spots that discharge higher than the desired level, since some of the pollutants would not be able to participate in the market. The initial allocation of permits has no impact on the efficiency of the market, but has potentially a big equity effect caused by the system selected for the allocation of the permits among the group of polluters. One important disadvantage, which is a general problem with efficient operation of markets, is the requirement to include many firms. If only a few firms trade, one may develop excessive influence in the market, leading to a market failure.

An example for functioning TDP in Australia is provided in the next section.

7.3   Hunter River Salinity Trading Scheme, New South Wales, Australia

Very few locations consider the option of a joint management of the nonpoint source pollution in a regional setting, using cooperative arrangements and trade, including use of permission permits among regional entities. The Murray–Darling Basin Authority (2001) initiated a basin-wide agreement, a joint work program designed for setting salt disposal permits on the basis of historical loads, including a revised cost-sharing formula and salinity credit allocation shares for Victoria, New South Wales, other South Australia states and the Commonwealth. In the Hunter Basin of New South Wales, Australia (Department of Environment and Conservation, 2003; NSW Minerals Council, 2014; NSW EPA, 2021), a scheme of salt permit discharges has been put into place. The main idea of this scheme is to permit discharge of salty water only when the assimilative capacity of the river allows it, namely when there is low salt and high quantities of freshwater in the river. Salinity disposals are restricted by quantity, time and location in the river.

Note

1.   Notice that economic considerations accept that society could be better off with a certain level of pollution, depending on social preferences and the relationship between abatement cost and damage from pollution.

References

Department of Environment and Conservation, New South Wales, Australia, 2003. Hunter River salinity trading scheme: Working together to protect river quality and sustain economic development. Accessed 26 February, 2022. https://www.cbd.int/financial/pes/australia-pesriver.pdf.

Mateo-Sagasta, J., S. M. Zadeh, H. Turral, and J. Burke, 2017. *Water Pollution from Agriculture: A Global Review*. Rome and Colombo: FAO and IWMI.

Murray–Darling Basin Authority, 2001. Basin salinity management strategy 2001–2015. Accessed 26 February, 2022. https://www.mdba.gov.au/sites/default/files/pubs/BSMS-full.pdf.

NSW EPA, 2021. Hunter river salinity trading scheme. Accessed 26 February, 2022. https://www.epa.nsw.gov.au/licensing-and-regulation/licensing/environment-protection-licences/emissions-trading/hunter-river-salinity-trading-scheme.

NSW Minerals Council, 2014. Review of Hunter River salinity trading scheme. http://www.nswmining.com/NSWMining/media/NSW-Mining/Publications/Submissions/140207_NSWMC-Submission_HRSTS-Review_FINAL.pdf.

Pacific Insitute, 2010. World water quality facts and statistics. Pacific Institute. Accessed 26 February, 2022. https://pacinst.org/wp-content/uploads/2013/02/water_quality_facts_and_stats3.pdf.

Schwarzenbach, R. P., T. Egli, T. B. Hofstetter, U. von Gunten, and B. Wehrli. 2010. Global water pollution and human health. *Annual Review of Environment and Resources*, 35:109–36.

UNEP (United Nations Environment Programme), 2016. *A Snapshot of the World's Water Quality: Towards a Global Assessment.* Nairobi, Kenya: United Nations Environment Programme.

8 Economics and politics of international water management

International water is a body of water that is shared by two or more sovereign riparian countries. Sometimes the term used is *transboundary water*, which refers to shared water among states in a federal system (Katz and Moore, 2011). The term *international water* includes shared water running in a river and water stored in an aquifer or in a lake. International water in rivers (e.g., the Euphrates-Tigris shared between Turkey, Syria, and Iraq) is different from water in aquifers (e.g., the Guarani Aquifer in South America, shared by Argentina, Brazil, Paraguay and Uruguay) or in lakes (e.g., Aral Sea, shared now by Afghanistan, Tajikistan, Kyrgyzstan, Turkmenistan, Uzbekistan and Kazakhstan). Water in lakes and aquifers is confined without significant natural move. In addition, shared water stored in aquifers is 'hidden' and hard to regulate.

Why is it important to discuss international water in this book? Management of shared international water affects the management of the remaining water resources of a given riparian country. The larger the share of a country's water resources coming from international water the lower is its ability to make long-term investment decisions in water infrastructure, and the more complicated it may become for that country to integrate the international water it is sharing into the national water grid. However, if the riparian states to an international water source (we will focus in this chapter mainly on river basins) have reached agreements on how to jointly manage the shared water, they are able to better also manage their 'own' water. In addition, since a significant portion of the world land surface is in shared international river basins, the study of management of international water is of great importance.

In what follows we discuss the extent of international water, review tools to manage international water, highlight differences between management of 'domestic' water and international water and suggesting economic concepts to allow better management of international water.

8.1   Basic information on international river basins around the world

It is estimated that globally, about 47 per cent of the land surface is part of international river basins (McCracken and Wolf, 2019). The borders of an international river basin and its area are subject to changes over time due to several effects, such as creation of new states (e.g., South Sudan) and/or occupation of one state, or part of a state, by another. Changes to cartographic capacity (use of satellite imaging) explain also possible changes to area and borders of international basins (Gambia river basin; see McCracken and Wolf, 2019: Fig 3). At present there have been 310 international river basins identified and delineated, which reflect changes to political boundaries and increased cartographic capacity over time (McCracken and Wolf, 2019). These river basins are shared by 150 countries and territories. Just as a comparison, the first attempt to register international river basins by United Nations Centre for Natural Resources, Energy and Transport (UNCNRET) of the Department of Economic and Social Affairs in 1978 identified only 214 international river basins.

Of the 310 international river basins registered at present, the majority (232) are bilateral, engaging two riparian countries only. The rest include 43 trilateral basins, 30 basins with 4–8 riparian states, one basin with 9 riparian states, one basin with 10 riparian states, one basin with 11 riparian states, one basin with 14 riparian states and one basin with 19 riparian states.

Management of shared international water has several features not found in water management by individual decision-makers. First, the decision-maker in international river basins are states and thus we may expect strategic behavior of the riparian states to be much more prevalent. Therefore, we can see much less policy intervention to regulate management of international water that impose prices or quotas and standards on the riparian states. Instead, what drives the management of international water are agreements between the riparian states. These agreements, also called treaties, are the result of negotiations between the riparian states. They follow to some extent the rules set in the global convention (on the Law of the Non-Navigational Uses of International Watercourses) signed by many (but not all) of the world countries (United Nations, 1997). It should also be noted that there are also international river basins with no treaty among all basin riparian states, and we can see also international river basins with agreements between some of the riparian states to the basin. An economic and political explanation to the difficulty to sign basin-wide agreements among international basin riparian states can be found in Dinar et al. (2019). It should also be highlighted that while it is less likely that international river basins with a large number of riparian states will sign a basin-wide agreement, the Danube, the international river basin with the largest number of riparian states (19), has a basin-wide agreement.

Management of international water often includes the sharing of water flows between the riparian states. But treaties also include other joint issues such as water quality (pollution) considerations in the basin, compensation for flood damages and sharing of the cost of joint investments in shared facilities in the basin (Dinar and Dinar, 2003). Globalization and international trade agreements among the basin states are expected to increase the level of cooperation by providing greater opportunities for policy coordination between the involved parties. Trade is likely to provide opportunities for linkage between environmental and trade exchanges, support implicit side payments, grant countries direct leverage over other countries' production and instill a perception of shared goals between countries (Sigman and Chang, 2010). Data reveal that water pollution is lower in rivers shared between countries with more trade, strengthening the view that trade promotes coordination of environmental policies. For more insights on the various angles from which international water management can be done, see Earle et al. (2010).

8.2   Economic and political considerations in managing international water

An important aspect in the management of international water is the geography of the river. There are about 13 geographies (Dinar, 2008) of river flows across and between the riparian states. However, there are two particular geographies that have been identified as affecting the political and economic interactions among the riparian states. The most important is the cross-border geography, defined as the geography where the river crosses from one country to the other. In such geography there are upstream and downstream states, which opens the door for exercising power and unilateral negative externalities that may affect the joint management of the shared water. The other important geography is the border-creator geography, where the river is the border between states. In this geography the unilateral negative externalities are less prominent and may provide more incentives to joint management of the river and its resources.

Under these geographies we may see that management of international water can take a non-cooperative form, where states along the water body take advantage of their strategic location or assets (Dombrowsky, 2007) to maximize their individual benefit, or a cooperative form, where the riparian states maximize the basin's benefit (Dinar and Hogarth, 2015). In the remainder of the chapter, we focus on the cooperative form of international water management. In making such selection, we assume that rational states should prefer cooperation to non-cooperation to maximize their outcome. This assumption is supported by several observations.

One observation is that water can easily be polluted with consequences that in most cases are unidirectional – affecting downstream, or neighboring, riparian states. Recognizing this phenomenon suggests that cooperation would minimize damage. In addition, water can be stored behind a dam or in an aquifer and be used for the benefits of several/all riparian states that share it. Polluted water can be treated and returned to the water system and be used by downstream riparian states. And finally, many water projects exhibit strong economies of scale, motivating collaboration to reduce average costs per unit of water. All these observations suggest that the wider the cooperation of the riparian states the larger the potential benefits from cooperation that could be shared.

So far, we assumed rational behavior of states, negligible negotiation and treaty implementation costs, and non-water-related political interests of the riparian states. While ignoring these aspects helps facilitate the analysis of the water and benefit-sharing problem, it leads to less realistic arrangements. Here is where the international relations and political science disciplines help economics with inclusion of political considerations that help reach more realistic management arrangements (Dinar et al., 2014; Dinar and Wolf, 1997). In the following sections we present one possible economic framework – the social planner allocation – that is used in international river basin water management to achieve economic efficiency and political feasibility.

8.3   The social planner model

The common approach to manage international water resources is to assign all basin resources to a supra-planner, called a *social planner*. The social planner 'ignores' political borders and considers only the optimization of the resources used across the basin. In other words, the social planner allocation of international water resources aims to maximize the joint benefit of all riparian states, subject to physical and institutional constraints relevant to the situation in the basin. An optimal social planner allocation is considered first best and serves as reference (benchmark) to which other allocations are compared. Deviations from the social planner outcomes represent inefficiency (benefit loss) of other alternative allocations.

The social planner allocation solution, while achieving the highest level of basin benefits, may not be acceptable to all riparian states compared to their level of benefits before the social planner allocation. It could be that the social planner allocation assigned water resources (or net revenues from water resources) at a lower level than a particular state had. Therefore, a mechanism for adjustment must be introduced. One possible mechanism is the introduction of trade in the social planner assignments among the states. This will enable riparian states to improve benefits relative to the initial (status quo) allocation and may reach the social planner allocation (Nigatu, 2012). Below are several cooperative game theory principles that allow states in the basin to participate in a basin-wide agreement that is efficient, stable and sustainable.

8.3.1   Allocating the social planner's basin-wide benefits to the riparian states

Once the social planner's solution is achieved, the surplus benefit it produces needs to be redistributed among the riparian states in an agreeable manner to all states. This means that the following requirements must be satisfied: (1) each state will get re-distributed benefits that exceed its benefits prior to the social planner solution – this is called the *individual rationality condition*; (2) the redistributed benefits to any combination of states in the basin is greater than any other allocation attained under any sub-coalition the state could establish – this is called the *group rationality condition*; and (3) the sum of the redistributed benefits to all basin riparian states (the grand coalition) fully allocates the total basin-level benefits (obtained by the social planner) – this is called the *efficiency condition*. Any redistribution of the social planner benefits that fulfills these three conditions is called the *Core*. The Core describes best the characteristics of the distribution of such incremental surplus, and leads to a stable, acceptable and efficient basin-wide allocation. Below is a demonstration of a solution in the case of the Blue Nile (Dinar and Nigatu, 2013).

8.3.2   The Blue Nile example

The Blue Nile river, part of the Nile Basil, is shared by Ethiopia, Sudan and Egypt. Nigatu (2012) developed a social planner model to demonstrate the usefulness of tradeable water rights, instead of the 1959 treaty-based allocation of the Nile basin to these three riparian states. In an extension, dealing with the distributional effects of possible cooperative solutions to the water allocation in the Nile, Dinar and Nigatu (2013) calculated the resulting allocations that are obtained once the institution of water trade among the states is allowed. Table 8.1 below presents the redistribution to all three basin states, under a given water right allocation (WRA2) scenario (see Dinar and Nigatu, 2013 for details).

The results in Table 8.1 suggest that the social planner solution for the basin created a total benefit sum of \$9.63 billion. That benefit sum is the highest that could be obtained. As we can see, under the institutional scenario WRA2, the basin is able to obtain \$9.56 billion, which is 0.997 of the social planner level. Redistributing this sum while reaching individual rationality, group rationality and efficiency conditions yields the allocation of benefits among all three states, which will keep them

ECONOMICS AND POLITICS OF INTERNATIONAL WATER MANAGEMENT 73

Table 8.1 Title benefits accrued by various coalitions in the Blue Nile Basin under various coalitional arrangements, water trade/no trade, abatement of soil erosion and allocation scenarios (billion USD in 2010 prices)

Allocation scenario	No water trade (status quo)			Water trade			
	Ethiopia	Sudan	Egypt	Ethiopia Sudan	Ethiopia Egypt	Sudan Egypt	Ethiopia Sudan Egypt
Social planner							9.63
WRA2	2.49	2.35	4.06	5.10 (2.90; 2.20)[a]	7.09 (2.63; 4.46)	7.08 (4.29; 2.79)	9.56 (2.76; 2.44; 4.38)

Note: [a] In parenthesis are the benefits that were accrued to each state in a coalitional arrangement.
Source: Adapted from Dinar and Nigatu (2013).

in a basin-wide agreement. Specifically, Ethiopia will get $2.76 billion, Sudan will get $2.44 billion and Egypt will get $4.38 billion.

Looking at actual cooperation among riparian states in various international water basins suggests the need to expand the definition of cooperation of riparian states of international water basins beyond only water.

8.4   Measuring actual cooperation in international water basins

So far, we have limited our analysis to certain characteristics that lead to agreement (cooperation) in managing a basin's water among riparian states of an international basin. We also demonstrated possible agreements in the Blue Nile case regarding flow and benefit allocations. However, riparian states also inflict pollution and ecological externalities in the case of a border-creator geography, which are not necessarily addressed by flow allocation or side payments (benefit transfers) among the states. In addition, we never defined what agreement/cooperation means beyond the assumption that it is motivated by certain behavioral traits of the participants (core conditions).

In the case of international water, the participants are nations (states) and not individuals. Therefore, several assumptions must be made regarding the preferences of a state and interactions among states. While these aspects are more prevalent in the field of international relations or political science, they are relevant as background in economic analysis of interactions among states in international water management.

Recent work on cooperation in international river basins has focused on the use of international water treaties as a measure of cooperation. International water treaties are agreements that specify rules for allocating the shared water. Works by economists (e.g., Ambec et al., 2013; Ansink and Ruijs, 2008; Van den Brink et al., 2012) and political scientists (e.g., Mitchell and Zawahri, 2015; Zawahri et al., 2014) address issues of efficiency, fairness and stability. While an agreement signals cooperation, it may well be an inefficient one and requires support, especially with the need to renegotiate it every several years (Brochmann, 2012). However, as argued by Zawahri et al. (2014), international water agreements are the basis for even wider and deeper cooperation beyond water. The fact that agreements are renewed indicates a stronger level of interest to cooperate that recognizes the need to renegotiate the agreement when conditions change.

8.5   Final comments

Management of water shared by several stakeholders (states) who are not part of one governing system poses challenges beyond economic considerations, and the literature on international water management has integrated multidisciplinary efforts to account for the strategic behavior of riparian states. Despite the multifaceted challenge of managing international water, experience so far suggests that economic principles and governing frameworks allow for acceptable and stable water management arrangements among riparian states. Indeed, use of water allocation concepts, such as water trade, in addressing scarcity issues that vary across the basin's parties are harder to implement in international water systems. However, the large potential benefits from joint management provide a strong incentive for collaboration. The results of the analysis in the Blue Nile, discussed above, suggest that the riparian states can achieve close to 100 per cent of the benefits resulting from the social planner allocation (Table 8.1). The literature offers additional factors that promote international collaboration in managing international water (e.g., issue linkage, as discussed in Pham Do and Dinar, 2014).

References

Ambec, S., A. Dinar, and D. McKinney, 2013. Water sharing agreements sustainable to reduced flows. *Journal of Environmental Economics and Management*, 66(3):639–55.

Ansink, E., and A. Ruijs, 2008. Climate change and the stability of water allocation agreements. *Environmental and Resource Economics*, 41(2):249–66.

Brochmann, M. 2012. Signing river treaties: Does it improve cooperation? *International Interactions*, 38:141–163.

Dinar, A., and M. Hogarth, 2015. Game theory and water resources: Critical review of its contributions, progress and remaining challenges. *Foundations & Trends in Microeconomics*, 11(1–2):1–139.

Dinar, A., and G. Nigatu, 2013. Distributional considerations of international water resources under externality: The case of Ethiopia, Sudan and Egypt on the Blue Nile. *Water Resources and Economics*, 2–3:1–16.

Dinar, A., and A. T. Wolf, 1997. Economic and political considerations in regional cooperation models. *Agricultural and Resource Economics Review*, 26(1):7–22.

Dinar, A., L. De Stefano, G. Nigatu, and N. Zawahri, 2019. Why are there so few basin-wide treaties? Economics and politics of coalition formation in multilateral international river basins. *Water International*, 44(4):463–85.

Dinar, S. 2008. *International Water Treaties: Negotiation and Cooperation Along Transboundary Rivers*. New York: Routledge.

Dinar, S., and A. Dinar, 2003. Recent developments in the literature on conflict and cooperation in international shared water. *Natural Resources Journal*, 43(4):1217–87.

Dinar, S., D. Katz, L. De Stefano, and B. Blankespoor, 2014. *Climate Change, Conflict, and Cooperation: Global Analysis of the Resilience of International River Treaties to Increased Water Variability*. Policy Research Working Paper No. 6916. World Bank, Washington, DC.

Dombrowsky, I., 2007. *Conflict, Cooperation and Institutions in International Water Management*. Cheltenham, UK and Northampton, MA, USA: Edward Elgar Publishing.

Earle A., A. J. Jagerskog and J. Ojendal (eds.), 2010. *Transboundary Water Management: Principles and Practice*. London: Earthscan.

Katz, D. L. and M. R. Moore, 2011. Dividing the waters: An empirical analysis of interstate compactallocation of transboundary rivers. *Water Resources Research*, 47:W06513.

McCracken, M. and A. T. Wolf, 2019. Updating the register of international river basins of the world. *International Journal of Water Resources Development*, 35(4):732–82.

Mitchell, S. and N. A. Zawahri, 2015. The effectiveness of treaty design in addressing water disputes. *Journal of Peace Research*, 52(2):187–200.

Nigatu, S. G., 2012. *Essays on Resource Allocation and Management, Price Volatility and Applied Nonparametrics*. PhD dissertation, Department of Economics, University of California, Riverside.

Sigman, H. and H. F. Chang, 2010. Implications of globalization and trade for water quality in transboundary rivers. In C. Ringler, A. Biswas and S. Cline (eds.), *Global Change: Impacts on Water and Food Security*, 97–111. New York: Springer.

Pham Do, K. H. and A. Dinar, 2014. The role of issue linkage in managing noncooperative basins: The case of the Mekong. *Natural Resource Modeling*, 27(4):492–518.

United Nations, 1997. Convention on the law of the non-navigational uses of international watercourses. Accessed 21 January, 2022. https://legal.un.org/ilc/texts/instruments/english/conventions/8_3_1997.pdf.

Van den Brink, R., G. van der Laan, and N. Moes, 2012. Fair agreements for sharing international rivers with multiple springs and externalities. *Journal of Environmental Economics and Management*, 63(3):388–403.

Zawahri, N., A. Dinar, and G. Nigatu, 2014. Governing international fresh-water resources: An analysis of treaty design. *International Environmental Agreements: Politics, Law and Economics,* 16(2):307–31.

9 Climate change and water resources

This chapter addresses the recent public discourse on possible impacts of climate change on water resources through a modified hydrological cycle (precipitation levels, distribution, evaporation and others) and policy interventions to address it. The ultimate result of climate change impacts on water resources is the direct and indirect impact on human activity, such as irrigated agriculture, hydropower production, ecosystem well-being and other negative effects. Once climate change impacts on the hydrological cycle are understood, scenarios of impacts and their distributional effects can be drawn and adaptation or mitigation policies can be designed and evaluated, using economic principles (Markandya, 2017).

In this chapter we start first with understanding the physical relationships between various aspects of climate change and various water parameters. Next, I describe the negative impacts of climate change on various sectors in the economy. Then we introduce individual and institutional responses in coping with climate change induced effects. I am aware of the many options to address climate change effects, but will not be able to cover all of them in this chapter. I will provide additional reading for those interested. A good reading to start with is UNESCO and UN-Water (2020), where general and non-technical aspects associated with the nexus between climate change and water are explained and demonstrated.

9.1 Impact of climate change on the hydrological cycle

Altered climatic conditions that are the result of increase in atmospheric CO_2 and other greenhouse gasses concentrations will most likely affect the distribution of precipitation both intertemporally and spatially, the quality of water, the resiliency of water supply systems, and human, animal and crop requirements for water resources in different regions. Climate change will likely lead to changes in soil moisture. Increased heat due to climate change will lead to higher levels of evaporation, leaving less water in the soil. It may also lead to higher level of evapotranspiration, reducing the efficiency of water use by plants. But the increase is expected to be partly offset by reduced plant water use in the CO_2-rich atmosphere. Higher temperatures may also impact the transitional winter snow zones, leading to changes in timing of snow fall and snow melt. More winter precipitation would be in the form of rain instead of snow, thereby increasing winter season runoff and decreasing spring and summer snowmelt flows. In locations where the additional winter runoff cannot be stored because of flood control considerations or lack of adequate storage, a damage in usable supply would be the result, in addition to direct and indirect flood damages, such as loss of life, loss of infrastructure, loss of yields, loss of access to school and others (McCarthy et al., 2001).

Precipitation is the main cause of variability in the water balance both spatially and intertemporally. Climate change may affect precipitation in various ways. In certain regions we see a reduction in the mean precipitation over time and at the same time we witness an increase in the interannual variation in precipitation. Therefore, changes in precipitation have very important implications for hydrology and water resource management. Hydrological variability over time in a river basin is influenced by variations in precipitation over different time scales such as day, season and year. Flood frequency is affected by changes in the interannual variability in precipitation and by changes in short-term rainfall properties (such as storm rainfall intensity). The frequency of low flows is affected primarily by changes in the seasonal distribution of rainfall (Columbia University, 2019).

Another effect of climate change is the increase in temperature that is the result of the global greenhouse effect. As is explained below, warming and cooling of the atmosphere leads to extreme effects of high temperature that alters sectors such as agricultural production patterns, making certain locations unsuitable for production while other locations that have been less suitable for production will transform to become well-suited agricultural production regions (Mendelsohn and Dinar, 2009).

In the following sections I provide explanations of several climate change-related mechanisms that affect the water cycle.

9.1.1   Mechanisms affecting precipitation

When the above-normal warm and wet air cools down, it drops above-normal precipitation in the form of rain or snow. In addition, certain places may see large increase in the intensity and frequency of heavy precipitation events (Columbia University 2019).

9.1.2   Mechanisms affecting evaporation

Warmer air contains more moisture than cool air. Therefore, in a warmer world, the air will absorb more moisture from oceans, lakes, soil and plants, leaving drier conditions in these areas, and negatively affecting water availability for various purposes (e.g., drinking, irrigation, environmental flows). At the same time, higher air humidity will make future higher temperatures unbearable in some places, by blocking the cooling effects of humans' body sweat. The change in air temperatures and circulation patterns will also lead to change in location of precipitation fall. Therefore, we may face locations that are expected to get drier, while other locations are expected to get wetter (Columbia University, 2019).

9.1.3   Mechanisms affecting surface runoff and stream flow

The more frequent, aggressive and longer events of precipitation caused by warmer, wetter air end up frequently in flooding, which can of course negatively affect human daily lives (e.g., school interruption, increased health risks), damage homes and crops, pollute waterways and the environment and negatively affect the entire economy. More intense rainstorms will also increase surface runoff (following storms). This strong runoff may lead to erosion and also capture pollutants such as sediments, dirt, physical derbies and other undesirable stuff, flushing them into nearby water bodies such as streams and aquifers, contaminating sources of water for agricultural and drinking supply (Columbia University, 2019).

9.2   Effects of climate change on water resources and social welfare

Once the interactions between climate change and water resources are understood and integrated into one model framework, it can be used to estimate the impact of climate change on the global economy or be used at a smaller scale to estimate effects of climate change, and climate change induced water scarcity, on regional or local economies and sectors. Several examples are provided below.

Most of the estimated impacts of climate change on water resources and on different segments of the economy are obtained via models' runs. For example, a global water model is used by Alcamo et al. (2007) to analyze the impacts of climate change and socio-economic driving forces (derived from the A2 and B2 scenarios of IPCC) on future global water stress. Their work extends previous global water work by analyzing not only the impact of climate change and population growth on water stress, but also effects of additional forces such as income, electricity production and water-use efficiency, among others.

Another example of a global relationship among climate change, water and economic performance can be found in a recent report (World Bank, 2016), which suggests that water scarcity, exacerbated by climate change, could affect economic growth, lead to domestic and international migration and ignite conflict among citizens, groups and states. The results also suggest that many countries can curb the negative impacts of water scarcity by improving water allocations among sectors and regions, leading to more efficient water use among the sectoral and regional demand. Important findings suggest that water will become even scarcer in regions where it is currently abundant, such as South Asia, due to the combined effects of growing populations, rising incomes and expanding cities, and water scarcity will greatly worsen in regions where water is already in short supply, such as the Middle East and Sub-Saharan Africa. The most important finding in World Bank's report (2016) is that the regions mentioned above could realize their growth rates decline by as much as 6 per cent of gross domestic production (GDP) by 2050 due to water-related impacts on agriculture, health and incomes.

A similar set of finding was reached by Mendelson et al. (2005). Their study estimated that climate change will have serious distributional impacts across countries, grouped by income per capita, measured in terms of GDP. The study predicted that poor countries will face the majority of climate change-related damages, mainly due to their geographical location (which is already hot and humid) and the fact that agriculture is still the leading sector in their economy, providing income and employment but also being vulnerable to climate change impacts. This primary reason is beyond adaptation ability, wealth resources and technological progress of poor countries.

9.3   Case studies from the wastewater sector and the agricultural sector

In this section I will provide examples of impacts of and adaptation to climate change in the wastewater sector (using data from China) and in the irrigated agricultural sector (using data from Júcar River Basin in Spain). Both the agricultural sector and the wastewater treatment sectors are vulnerable to climate change and may face irreversible damages if water allocation, investment and long-time planning are not done properly. The two examples are taken from work I have done with my students in previous years.

9.3.1   The impacts of climate change on wastewater treatment costs

Treatment of wastewater is becoming a major developmental and environmental issue around the world and is expected to become an acute need in the years to come due to increases in population and level of water scarcity. Reznik et al. (2020) identified a direct relationship between climate and treatment costs of wastewater, which has manifested in making this water source sustainable for future use. Being a relatively stable source of supply wastewater can substitute scarce natural fresh water sources. But the costs of the treatment process, those associated with direct input use (in addition to regulatory costs), might hinder the attractiveness of this resource.

Climate change may affect the wastewater infrastructure through increasing uncertainty in future air temperature, precipitation, wind speed and rise in sea level, all of which may destroy such infrastructure. The wastewater sector is being affected by climate change in various ways, including higher amounts of pathogens from storm water collection systems connected to the wastewater facility following extreme rainfall events. The increased amount of pathogens and the extreme rainfall events may lead to operational needs above and beyond the treatment plant's designed capacity, affecting the reliability and operating costs of the facility. In addition, climate change, by raising temperature and humidity, affects the biological processes of treatment plants by reducing their efficiency and making them less effective.

Reznik et al. (2020), using data from 163 wastewater treatment plants representing all regions in China, were able to estimate the treatment costs of the wastewater sector in China, and simulate changes in these costs with expected future climate conditions, policy implementation scenarios, population growth and development trends. The estimation of the treatment cost function includes, as explanatory variables, climate indicators at the plant level (data on climate in the vicinity of the wastewater treatment plant). This enables capturing the impact of changes in climate on costs of treatment, but it also allows distinguishing, as part of the simulation, between predicted costs when climate changes are accounted for or ignored.

Different policy implementation scenarios were evaluated regarding their effectiveness in sustaining the treatment costs for various changes in the factors associated with wastewater production: (1) population growth, (2) per-capita water consumption, and (3) urbanization growth rates. The policy tested was treated wastewater discharge standards, requiring all plants in the dataset to treat wastewater to a given level as defined by the national standard for treated wastewater discharge. The policy was applied under different implementation scenarios that vary by future climate trajectories.

The analysis suggests that climate change impacts on wastewater treatment costs can be substantial. The estimated opportunity cost associated with ignoring these potential effects was also found quite significant. Depending on the discount rate, variation of the opportunity costs estimated among policy implementation scenarios, given uncertain future development, could be substantial as well. Yet, depending on predicted changes in climate from different general circulation models, in some cases climate effects will decrease or even reverse the predicted impacts resulting from well-informed estimation alone. For the former, the cost of ignoring climate change might be negligible. These cases, however, are the minority according to the results of their analysis.

In general, comparing across policy implementation scenarios, ignoring climate change impacts on future planning of wastewater treatment could have different estimated opportunity costs, suggesting that information on future climate change impacts could be critical for efficient policy design.

9.3.2 Adaptation to climate change in arid and semiarid agricultural regions

Climate change is projected to exacerbate water scarcity and increase the recurrence and intensity of droughts, especially in semiarid regions. Kahil et al. (2015) developed and applied a hydro-economic model that links a reduced form hydrological component of a river's natural flow and human intervention in storage and diversions, with economic and environmental components. They applied the model to the Jucar River Basin (JRB), which is a semiarid region in Southeastern Spain, to analyze the effects of droughts and to assess alternative adaptation policies. The model includes the following components: hydrology of the river and the interaction with groundwater aquifers, reservoirs and hydroelectric generation; irrigated agriculture operations; urban/residential consumption; and environmental needs. Having all these components in one integrated model allows a comprehensive analysis of natural shocks (climate change–induced droughts) and policy interventions.

Climate change impacts in the JRB indicate a reduction of water availability by 19 per cent in the short term (2010–2040), and 40–50 per cent in the long term (2070–2100). The drought scenarios considered in Kahil et al. (2015) cover the range of these estimations. Their model was used to evaluate the economic and environmental effects of policies to combat alternative climate change induced droughts. Three policy intervention alternatives were considered.

1. Baseline policy: The current water management in the JRB to cope with water scarcity and drought. The current policy allows flexible adaptive changes in water allocations, following negotiation and cooperation between the different user groups, taking into account the decision-making process and environmental concerns of the various parties.

2. Agricultural–urban water market: Market-based allocation of scarce water during droughts could allow water transfers between willing buyers and sellers, resulting in welfare gains to both. Under this policy, water trade is allowed among irrigation districts and between irrigation districts and urban users in the JRB.
3. Environmental water market: Similar to policy 2, recent years have emphasized the importance of water trade for securing water needed for the environment. A water trade policy with the goal of securing the environment includes the same agents that trade water under policy 2, but also a steward for the environment (that is not represented by any agent) in the form of a basin authority or any government agency that participates in the water trade system in order to secure environmental amenities. In the case of the JRB the subject of the concern of the basin authority is the Albufera wetlands.

The difference between policy 2 and policy 3 is that in the agricultural-urban water market the traders are private decision makers interested only in maximizing their own benefits, while in the environmental water market the objective is to maximize social benefits (including benefits from better environmental performance), and thus an intervention by a social (government) regulatory agency is needed.

The model, developed by Kahil et al. (2015), was used to compare the effects of each of the policies on the basin-level benefits under each climate change-induced water scarcity level. The results indicate that all drought levels investigated have considerable impacts on social benefits welfare, with the main adjustments sustained by irrigation and the environment sectors. The water market policy seems to be a suitable option to overcome the negative economic effects of climate-induced droughts, although under policy 2 it appears not sufficiently effective in terms of the benefits provided by the environment. This inefficiency has been taken care of in policy 3, where environmental needs have been addressed by the river basin authority. The results of the social welfare in the basin under each climate-induced drought scarcity show the following priority order: $3 > 2 > 1$.

9.4   Summary

Climate change affects social welfare via different channels. One important channel is the water system. Being central to economic activity, the water system can magnify or contain the impact of climate change on the economy and the individuals in the economy. In this chapter we discussed the various mechanisms by which climate change affects the water system, and the various adaptation options to sustain activity in the sector or the economy. Using two case studies from important water-related sectors, wastewater and irrigated agriculture, we further realized how proper policies can minimize the negative impacts of climate change on the performance of those sectors.

References

Alcamo, J., M. Flörke and M. Märker, 2007. Future long-term changes in global water resources driven by socio-economic and climatic changes. *Hydrological Sciences Journal*, 52(2):247–75.

Columbia University, 2019. How climate change impacts our water. *Columbia Climate School*, 23 September, 2019. https://news.climate.columbia.edu/2019/09/23/climate-change-impacts-water/.

Kahil, M. T., A Dinar, and J. Albiac, 2015. Modeling water scarcity and droughts for policy adaptation to climate change in arid and semiarid regions. *Journal of Hydrology*, 522:95–109, 2015.

Markandya, A., 2017. *State of Knowledge on Climate Change, Water, and Economics*. Water Global Practice Discussion Paper, World Bank, Washington, DC. https://openknowledge.worldbank.org/handle/10986/26491.

McCarthy, J. J., O. F. Canziani, N. A. Leary, D. J. Dokken, and K. S. White, 2001. *Climate Change 2001: Impacts, Adaptation, and Vulnerability: Contribution of Working Group II to the Third Assessment Report of the Intergovernmental Panel on Climate Change*. London: Cambridge University Press. https://www.ipcc.ch/site/assets/uploads/2018/03/WGII_TAR_full_report-2.pdf.

Mendelsohn, R. and A. Dinar, 2009. *Climate Change and Agriculture: An Economic Analysis of Global Impacts, Adaptation, and Distributional Effects*. Cheltenham, UK and Northampton, MA, USA: Edward Elgar Publishing.

Mendelsohn, R., A. Dinar, and L. Williams, 2005. The distributional impact of climate change on rich and poor countries. *Environment and Development Economics*, 11(2):159–78.

Reznik, A., Y. Jiang, and A. Dinar, 2020. The impacts of climate change on wastewater treatment costs: Evidence from the wastewater sector in China. *Water*, 12(11):3272.

UNESCO and UN-Water, 2020. *United Nations World Water Development Report 2020: Water and Climate Change*, Paris: UNESCO. https://unesdoc.unesco.org/ark:/48223/pf0000372985.locale=en

World Bank Group, 2016. *High and Dry: Climate Change, Water, and the Economy*. Washington DC: World Bank. https://openknowledge.worldbank.org/handle/10986/23665.

10 Emerging topics in water economics and policy

As water is recognized to be more and more central to the economy, and as water scarcity has been increasing over time in many locations, the importance of various water analyses and contexts becomes a critical aspect that must be considered when addressing water policies and water sector performance. We reviewed several important topics in water economics in the previous chapters. But we have left many others untouched. In recent years, we witnessed several topics in water economics and policy that gain their importance in the water economics literature and should be of interest to readers of this book.

This chapter introduces, in a very short manner, several special important issues in water economics and policy that were not covered in the previous chapters of the book. Topics include managed aquifer recharge, inter-basin and intra-basin water transfers, and the regional arrangements for use of treated wastewater for irrigation. All topics are intended to serve as a policy to address water scarcity. The important aspect of these topics is the way they have been analyzed – regional, multi-water resources, multi-sector approaches, all of which are preferred frameworks to analyze water issues and economic issues in a comprehensive manner. These three topics are based on my recent work with my graduate students and postdocs, and references are provided for those who are interested in more details.

10.1 Managed aquifer recharge[1]

Managed aquifer recharge (MAR), defined as intentional storage of water of various qualities in aquifers, is considered a potential strategy to mitigate water scarcity and negative drought effects. Below, one can find a description of a suggested framework to evaluate the economic performance of specific policies and institutions under a set of climate change scenarios, applied to the Kings Groundwater Basin in California's Central Valley, as an example.

MAR is a set of practices and institutions that allows the recharge of water of various types and qualities (surface water, recycled wastewater, and even groundwater from different locations) into a given aquifer. It therefore can reduce groundwater over-pumping-induced land sinking (subsidence) damages, prevent saline water intrusion, protect wetland habitat, provide flood protection, and more.

This section examines the role MAR can have for the Kings Groundwater Basin in the Central Valley of California, based on work by Reznik et al. (2021). Using several climate change scenarios, it evaluates how MAR performance is impacted by possible institutional arrangements and regulatory policy interventions.

10.1.1 The features of the analytical framework

The model used by Reznik et al. (2021) is based on an iterative process of two models: a hydrologic model (WEAP/CVPAM) and a dynamic economic optimization model (EOM). The hydrologic model represents in detail the Kings Groundwater Basin and surrounding area. The EOM combines an optimization model of the water and agricultural sectors – calibrated and applied to the same region, where net revenue of farm-level operation is sought. The important features of the modeling framework are the regional setup – including all sub-basins that are part of the Kings Groundwater Basin and including all types of water available in the region (surface water, groundwater, wastewater).

Reznik et al. (2021) identified seven Decision Analysis Units (DAUs) in the region, which serve as decision-makers. These DAUs represent water/irrigation districts, or urban centers (in this case it is the city of Fresno). Twenty different land use categories (e.g., crops) in the region, including land fallowing, were modeled. A production function was matched to all grown crops in the region. The EOM was calibrated to actual water prices, land allocations, and local water quota conditions in the region. At the end of the iteration process, the EOM produces an optimal set of various types of water uses (combination of surface water, groundwater, and recycled wastewater), land uses, and water quantities to be pumped from and recharged to the aquifer, and of course the derived net benefits for the sub-regions in the Kings Groundwater Basin.

10.1.2 Set of Policy Scenarios

The hydrology-economics analytical framework described above was used to evaluate three alternative policy scenario interventions. The first, which sets the benchmark for the other two, is the social planner solution (termed *Social*).[2] The second, termed *Sustainable*, is constructed based on the California Sustainable Groundwater Management Act of 2014 (SGMA). Under this scenario, it was required that for each subdistrict the groundwater head at the end of the 20-year planning horizon will be greater or equal to its baseline initial level at the onset of the planning horizon. The third policy scenario, termed *Credit*, implements the institutional well-known principles of 'capacity sharing' of an aquifer system (Dudley and Musgrave, 1988). According to this policy, each DAU holds a credit, based on the storage capacity of the aquifer, which limits the annual amount that can be extracted from the storage/aquifer. An initial endowment of annual credit is assigned to each DAU, which increases with intentional recharge (through infiltration basins) to the aquifer and decreases with pumping throughout the planning horizon. The limitation on groundwater extraction is unique to the *Credit* scenario. In the other two scenarios, groundwater extraction is limited only by hydrogeologic feasibility constraints, preventing groundwater level to fall below some minimal threshold that is set exogenously (by the social planner).

10.1.3 Climate scenarios

For each of the policy scenarios described, three climate scenarios were simulated in terms of regional rainfall and surface water availability from the Kings River, the San Joaquin River, and the Friant-Kern Canal. Under the first simulation, termed *Average*, Reznik et al. (2021) assumed constant values of climate conditions throughout the entire planning horizon, set at the long-term annual means of rainfall and surface water availability in the region. The second, termed *Hist1*, assumes regional climate conditions similar to those in the period 1975–96. The third, termed *Hist2*, refers to the climate conditions in the region during the period 1983–2004. *Hist1* and *Hist2* represent two climatic patterns in the region – relatively wet and relatively dry, respectively. The three climate scenarios differ in their precipitation and flow in the different rivers that cross the region. The models were run for a period of 20 years.

10.1.4 Main results

In this section I review very briefly the general effects of the different parameters without getting into the detailed distribution among the different DAUs. Interested readers can reach out to Reznik et al. (2021).

It was found that the total annual agricultural water use in the region declines from about 1.4 million acre feet (MAF) to about 1.2 MAF over the planning horizon of 20 years (1 acre foot = 1,235 cubic meters). Examining water use trends at the sub-region and crop levels indicates major differences regarding crops grown, and amount of irrigation water applied, between the DAUs. The time-paths of water application levels, averaged for each DAU, suggest that deficit irrigation is found optimal across most of the region's DAUs and throughout most of the planning horizon.[3] Lower water application levels are attributed to a decrease in regional groundwater extractions, compared to observed values, which according to the model results is increasing from 150 thousand acre feet (TAF) per year to about 200 TAF per year, throughout the planning horizon. It was also found that recharge of groundwater is achieved through excess irrigation of field crops and from intentional percolation of treated wastewater in certain DAUs.

10.1.5 Policy and climate scenarios

Comparing land allocation results across policy scenarios and under *Average* climate conditions suggests that the results of the *Sustainable* scenario are similar to the results of the *Social* scenario. However, the results of the *Credit* scenario suggest a dramatic decrease in groundwater extraction and total water use in the region, and consequently an increase in land fallowing, mostly at the expense of tree crops, and to a smaller extent of field crops in specific DAUs. For the rest of the region, land allocation differences compared to the other policy scenarios are far less significant.

Under the *Sustainable* scenario, total water use in agriculture is higher than in the *Social* scenario. Similar to the *Social* scenario, some DAUs demonstrate a decreasing trend in water use. However, towards the end of the planning horizon water use increases again. This translates into excess irrigation, mainly at the beginning and the end of the planning horizon, which is manifested in significantly higher recharged quantities to groundwater than under the *Social* scenario. The *Sustainable* recharge strategy results in higher regional groundwater levels, on average, compared to the *Social* scenario. As in the *Social* scenario, intentional recharge through infiltration basins is not preferred.

Results of the *Credit* scenario suggest a substantially lower use of water in agriculture compared to the *Social* scenario. Groundwater pumping is profoundly lower, and reused quantities of treated wastewater are higher under this scenario compared to results of the *Social* scenario. Due to lower water use in agriculture, recharged quantities are also smaller for this scenario compared to the other scenarios. However, as mentioned, pumping is also considerably smaller; therefore, regional groundwater level increases on average over time, more than in the *Social* scenario. Different to the other scenarios, intentional recharge through infiltration basins is found optimal under the *Credit* scenario. This is because some subdistricts in some DAUs rely solely on groundwater, which forces recharge as a means to accumulate credit to enable groundwater extraction throughout the planning horizon.

Comparing results between climate simulations and under the *Social* scenario, it was found that treated wastewater and groundwater storage are used as sources for stabilizing supply and smoothing consumption. This is when significant reductions in surface water supply occur under the *Hist1* and *Hist2* climate simulations. A second interesting result from the analysis of climate simulations emerges when comparing land allocation across all scenarios. For the most part, land allocation to crops in the region remains similar regardless of the assumed policy scenario or climate simulation used. The exception to this rule is the *Credit* scenario, in which land allocation results are sensitive to the assumed climate conditions.

Table 10.1 presents total regional economic welfare differences compared to the *Social* scenario in annual terms, across policy scenarios and climate simulations. The institutional arrangement under the *Credit* scenario inflicts significant implications on the region, resulting in a detrimental welfare impact amounting to about $2 billion annually. As noted earlier, for this scenario, the differences in the optimal plan across climate simulations are substantial, resulting in a significant difference in the

Table 10.1   Reductions in economic welfare compared to the *Social* scenario benchmark ($1,000 USD)

	Average	Hist1	Hist2
Sustainable	7,893	9,921	9,879
Credit	1,756,849	2,283,592	1,732,795

Note: The regional economic welfare achieved under the *Social* scenario is the highest. Therefore, differences presented in Table 2 are all negative.
Source: Adapted from Reznik et al. (2021).

economic welfare of about $500 million annually. The economic cost of the *Sustainable* scenario is in the range of $8–10 million annually, which is relatively mild.

The difference in regional economic welfare divided by total recharged quantities is used as an upper limit of the dollar value for recharged quantities in the region. Table 10.2 presents the differences compared to the *Average* climate simulation of total recharged quantities in the region over the entire planning horizon, as well as the differences in economic welfare per unit of water recharged across policy scenarios for the climate simulation *Hist1* and *Hist2*.

Total quantities recharged in the region are lower for both climate conditions *Hist1* and *Hist2* compared to *Average* conditions. The value of an acre-foot recharged in the region is in the range of $38 to $1,556 USD. Excluding the *Credit* scenario under *Hist1* climate conditions, the value of unit of water recharged exceeds the value of water in production, which is in the range from $50 to $250 per acre foot (AF). This suggests that the indirect benefits associated with recharged water quantities are substantial and surpass their direct benefits in most cases. Thus, the results suggest that recharging groundwater in order to support the optimal plan under different institutional arrangements is of high value to the region studied.

10.1.6 Summary and policy implications

The results of the analysis, according to the first-best scenario, suggest a significant reduction in groundwater use, which is complemented by deficit irrigation, without inflicting significant changes compared to observed crop-yield levels and land-use decisions. Excess irrigation of field crops and some flooding of fallowed land at the beginning of the planning horizon is the preferred method for recharging groundwater stocks, regardless of the assumed climate conditions. This strategy is amplified when minimal threshold levels of groundwater head at the end of the planning horizon are imposed, as part of our *Sustainable* scenario, suggesting that this institution incentivizes intentional recharge and increases its value to the region. Diverting water to infiltration basins and away from irrigation of crops is only warranted under the *Credit* scenario, and at a high economic cost, suggesting there are substantial trade-offs associated with this recharge method for the region.

Institutional impact is substantial according to the results of the analysis. This is demonstrated through the changes in regional land allocation and water use decisions under the *Credit* scenario, causing detrimental economic implications compared to the *Social* benchmark. Results from the analysis also suggest that the impact of future climate uncertainty on the region is highly dependent on the prevailing institutions and provides an estimated $500 million USD annually as an upper limit for the regional economic damage associated with uncertainty in water availability.

Table 10.2 Differences in total recharged quantity and economic welfare per unit of water recharged compared to the *Average* climate simulation and across policy scenario

	Differences in total quantity recharged (TAF)	Differences in economic welfare per unit of water recharged ($/AF)
Hist1		
Social	667	1556
Sustainable	447	421
Credit	713	38
Average	609	671
Hist2		
Social	316	682
Sustainable	323	294
Credit	116	332
Average	251	436

Source: Adapted from Reznik et al. (2021).

The *Sustainable* scenario presents a good compromise for the region between the ideal benchmark (*Social*) and the more stringent institutional arrangement (*Credit*). In the *Sustainable* scenario, groundwater levels increase the most, economic losses are small, and the simulated climate conditions appear to have a minor impact on the optimal strategy. This in turn implies that this institution is likely feasible and relatively easy to implement, monitor and enforce.

10.2   Inter-basin and intra-basin water transfers as a policy to address water scarcity[4]

Water scarcity levels vary across geographical locations and over time. Assessments of spatial and intertemporal water scarcity levels (WWAP, 2019) suggest that by 2050 economic growth and population change alone could lead to an additional 1.8 billion people living under various levels of water stress,[5] with the majority of them in developing countries.

Clearly, supply of water in many parts of the world lags behind the demand under existing conditions and the actual use by sectors. The substantial reduction (including deterioration in water quality) in the available renewable water resources over time and the increase in water-consuming economic activities (e.g., food production, increase in standards of living, deterioration in water quality) has led to a widening gap between the water quantities supplied and demanded.

Water scarcity at the country, state, county, and city levels could be addressed by using different policy intervention mechanisms directed both at demand- and supply-side management. For supply-side management, which is the focus of this section, policies include public investment in water supply, including manufacturing of water from the ocean (desalination) and treating wastewater to potable conditions for human consumption, or treating wastewater for use in agriculture or for aquifer recharge (several policies of which have been discussed in previous chapters of this book).

One major supply-side management policy is the transfer of water from locations where it is relatively abundant to locations where it is relatively scarce, both within one river basin (intra-) or between discrete river basins (inter-). Inter- and intra-basin transfers are simply 'the transfer of water from one basin to another distinct basin or river catchment, or a sub-basin within a shared basin or river reach, respectively' (Gupta and van der Zaag, 2008: 29). This policy involves massive investments in infrastructure and storage and is different from a policy that allows the introduction of water trade for marketing of water among water rights holders, using existing conveyance infrastructure.

Intra- and inter-basin transfers (IBTs) have been practiced and reported in the literature. IBT affects the local population and the environment in water-exporting regions, water-transmitting regions (through which water moves from the exporting to the importing regions), and in water-importing regions. By transferring water from a natural habitat for the purpose of benefitting a different region, could the policy itself become harmful? The policy was established to move water, and the original intention was appropriate, but there could be unforeseen and obstructive outcomes. Some of the outcomes result in immediate problems, such as perpetuating inequality through the act of displacing people close to a basin or in the path of a pipeline or canal that is used for the purpose of transferring water. However, sometimes the damage takes a long time to surface or to be realized; for example, the selenium problem that occurred on the west side of the San Joaquin Valley surfaced in 1985, 25–30 years after the federal and state water transfer projects (CVP and SWP, respectively) had begun operation.

Purvis and Dinar (2019) used information in the published literature on IBTs to assess their sustainability as a policy intervention aimed to ease water scarcity at both the local (intra-basin transfer) and regional (inter-basin transfer) levels. Since very few quantitative cost–benefit analyses are available for IBTs, they applied QUAlitative Structural Approach for Ranking (QUASAR) to 121 studies of documented IBTs from around the world, using five attributes of impacts that were associated in the literature with the transfers. These attributes include social impacts (private impacts are included where applicable), including any negative and positive externalities that are found within the literature describing these IBTs.

10.2.1 Effects of IBT: Affected parties, interested parties, and benefits and losses

IBTs are associated with both positive and negative impacts to water-exporting, water-transmitting and water-importing regions. IBTs are associated with building reservoirs and dams. But the positive impacts upon individuals and groups brought by these dams have been offset by negative social, economic and environmental impacts, such as high numbers of displaced people (estimated to range from 40 to 80 million), lack of equity in the distribution of benefits and negative impacts on ecological services – rivers, watersheds and aquatic ecosystems.

IBTs are also associated with indirect costs in the form of political objection or persuasion of the population in the exporting and importing regions. For example, in 1992, Yemeni people living in Habir discovered that their region had been 'proposed as the next source of water' for Ta'izz city; however, the people living in Habir had seen the impact that a water transfer had on the neighbouring valley of Al-Haima, so they were determined to stop the transfer. They were able to postpone the project until 1995 but, eventually, a financial understanding was agreed upon, and the villagers allowed the construction to take place (Jagannathan et al., 2009: 251).

In some countries with very large land areas, the inter-basin water transfers are on the agenda of policy makers for quite some time. Mahabaleshwara and Nagabhushan (2014) review the challenges (engineering, political and institutional) associated with several of the river inter-links that have been proposed and initiated to a certain extent.

Negative impacts, including those in the exporting regions and basins, involve higher levels of aridity and increased levels of salinity, both of which damage the ecology of the exporting basins. Additional damages are associated with an increase in water consumption in the importing basins, and the possible spread of disease due to contamination of the water at the source or along the conveyance route. In addition, some effective alternative measures have been established for IBTs, such as improving efficient water use, investing in desalination technology and rainwater harvesting practices (Zhuang, 2016: 1). One of the major features of an IBT, which is commonly labelled as 'positive,' is the redistribution of water resources between regions. However, this could lead to inequity for the exporting basin and also the territories or areas between the exporting and importing basins, where the conveyance infrastructure is proposed to be constructed, showing that IBTs could be double-edged swords (Zhuang, 2016). These construction projects will most likely introduce lasting issues to the surrounding territories and to the ecological environment. Compared with water conservation measures (Rinaudo and Barraque, 2015), IBTs involve far more problems that could arise along a more distant horizon (Zhuang, 2016).

The environmental impacts of water transfers vary around the world. For example, in Bangladesh negative impacts on fish resulted from an IBT, and from an environmental justice perspective, displacement of people and loss of crops occurred (UNESCO, 1999: 84). In the Sibaral Basin in Central Asia, negative environmental justice issues resulted because of the river reversal where Central Asia and Kazakhstan called for water within the Ob River to be redirected south, where the water was actually needed (UNESCO, 1999: 95, 99). Another example is the Snowy Mountains in Australia, which were met with adverse impacts because of the scheme that produced gigawatts of electricity for the surrounding region, and also because of poor natural resource management, leading to reduced flows below each diversion point (UNESCO, 1999: 101).

Additional explanations by Yevjevich (2001) suggest that while it is generally assumed that water users in the exporting region will lose future benefits of the water by allowing for that diversion, water users in the importing region will benefit from importing that diversion. This creates a winner vs. loser scenario, in which the exporting region loses for having a surplus of water, and the importing region is rewarded for having a deficit. However, it is not that simple. It needs to be assessed and, if possible, enumerated how much damage will occur in the region that receives the diverted water, and if any positive impacts could occur in the region from which the water is diverted. These enumerations are called *benefit–cost analyses* (Boardman et al., 2018) and are a cornerstone of the analysis of major water diversion projects. Diversions or transfers of water are multi-disciplinary problems with a physical aspect that is related to the geomorphology, water quality and overall water resources in the exporting and importing regions. Hydraulic engineering assessments should determine how the water will actually be conveyed from point A to point B. For the sociological aspects of a diversion or transfer, the components are 'political, administrative, economic, ecological, environmental, and legal,' because these components define how people and the environment will be *affected* by this transfer of water (Yevjevich, 2001: 342). Inter-basin transfers 'require joint planning' between the interconnected basins. In more complex cases, though, several basins become interconnected and need to cooperate in order to implement the transfer successfully. In extreme cases, all river basins and regions through basin or regional water transfers become interconnected, such as the National River Linking Project in India (Yevjevich 2001: 343).

Snaddon et al. (1999) provide a qualitative accounting of the impacts of the 51 existing and 53 proposed IBTs in their study. While the impacts assessed are only described rather than measured, it gives us a meaningful grasp of what constitutes the range of impacts of IBTs. Snaddon et al. (1999) identified (but did not monetize or quantify) several types of impacts affecting part of the 104 IBT projects. Such impacts include positive and negative effects both in the exporting and in the importing regions. Certain IBTs could be affected by several of the following impacts: (1) effects on water quantity and flow patterns; (2) effects on erosion and geomorphology of riverine systems; (3) effects on groundwater resources; (4) effects on water quality; (5) effects on aquatic ecosystems; (6) socio-economic effects; and (7) cultural and aesthetic effects.

Purvis and Dinar's (2019) assessment model was developed as follows: They identified impact attributes that were discussed by the authors of the IBT projects reviewed. Each IBT reviewed was associated with up to a given number of attributes. Several analyses exist for the same IBT. Some of the analyses included subsets of the number of attributes. Therefore, in these cases, the attributes from the various reports were combined. Purvis and Dinar (2019) did not find different impacts (e.g., negative or positive effects) in the various publications addressing the same IBT. Because no quantitative measurement was provided, even in the few studies that included benefit–cost analyses, they just marked whether or not an attribute is mentioned as having a positive, negative or neutral impact.

Some of the sustainability attributes that can be identified in relation to the intra- or inter-basin transfer could remain neutral as a result of the transfer, some could be affected negatively, and some could be affected positively. In each particular reported transfer IBT study effects could be identified on all or a subset of the sustainability attributes. Therefore, where a given attribute was mentioned, but no particular cost or benefit or indication about the direction of the effect was mentioned in the reported analysis of the transfer, it was assumed that the impact of the transfer on that particular attribute is neutral (zero).

This transformation of the impacts on different attributes is similar to a Likert scale, because it has assigned values of –1, 0 and 1 in order to value this qualitative data in a quasi-quantitative manner. For more explanation on the Likert scale-like metric see Purvis and Dinar (2019). This method was chosen for the purpose of the analysis because of its ability to cater to either extreme, or remain conservative (close to center, the position of neutrality). The attribute values were assigned based on how a reviewed study described an attribute related to a transfer: whether the transfer was 'damaging' or 'beneficial' for humans or the environment was used as the basis of coding each respective attribute.

During the review of the identified IBT cases in the literature, Purvis and Dinar (2019) marked the various impacts identified by the authors of each IBT analysis/description. Impacts of IBTs were divided into five categories/themes: (1) efficiency management effect (EM), measuring the overall efficiency associated with the management of the basin water; (2) irrigation outcome effect (IR), measuring the effects associated with irrigation projects, which were the destination of the transferred water in many cases; (3) environmental rehabilitation effect (ER), measuring the effects of transfers aimed to move water in order to support ecosystems in the receiving regions; (4) environmental/ecological effect (EE), measuring the pervasive effects of the transfer on the environment/ecosystems; and (5) environmental justice/equity effect (EJ), measuring the social and cultural effects of the transfer on indigenous and poor communities in the exporting, transmitting and importing regions.[6]

The total net impact of an IBT can also be presented using an area graph (Dinar and Saleth, 2005) as is depicted in Figure 10.1 for one IBT with relatively overall high impacts in the sample.

10.2.2 Discussion

Facing various water-scarcity levels and their impacts, policymakers consider IBTS among other policy interventions such as pricing of water and local supply amendment projects. While many of the other policy interventions have been evaluated for their effectiveness and efficiency, IBTs, which are much more complicated interventions, have not (at

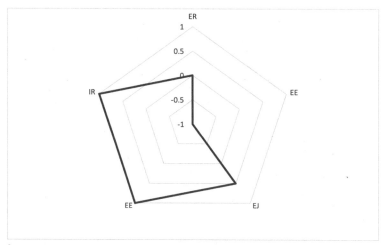

Source: Modified from Purvis and Dinar (2019).

Figure 10.1   Area graph representing the impact of five attributes of
the Istranca and Konya plain projects in Turkey

least in the global context), mainly due to the difficulties to harmonize
the analyses across all cases and to find appropriate economic indicators
for all possible attributes/impactS of the transfers. In recognition of such
difficult conditions for a global analysis, and with the growing importance
of comparing across the many IBTs reported in the literature, Purvis and
Dinar (2019) apply a simple, but useful, approach that allows the dichot-
omous identification of impacts associated with various attributes/effects
of the transfers.

The distributions of the five attributes of the IBT effects that were identified in the 121 IBTs reviewed by Purvis and Dinar (2019) show that
several attributes are harder to obtain in the positive range than others.
Environmental or ecological effects (EE), irrigation outcome effects (IR)
and environmental justice/equity effects (EJ) are associated with negative
means, indicating the difficulty to obtain positive effects of these attributes in a water transfer. All other attributes (EM and ER) are associated
with a small positive effect.

Nearly half of the transfers that target irrigation projects (IR) were found
to be associated with negative impacts, which is bothersome due to the

fact that irrigation water captures 60–90 per cent of the available water in the world. Therefore, the irrigation component might have a much larger negative effect if the impact can be calculated using more specific impact analysis methods. IBTs that include efficiency management effects (EM) and environmental rehabilitation effects (ER) have scored a lower percentage of negative effects in total (around one-third of such IBTs had negative effects). Overall, it could be seen that there is a negative share for the assigned codes (if a zero effect is not included within the positive effects) with nearly 70 per cent of the IBTs with an overall negative score.

10.2.3 Conclusion, policy implications, and caveats

Regional water transfers, whether inter- or intra-basin, deal with or address scarcity by acting as a means for protecting the water security within several basins. Since there is a problem with equitable and efficient allocations, there inevitably will be an inequitable distribution of water externalities. In some regions, environmental issues such as land subsidence exist in exporting regions, or the exporting or importing regions may be at risk of losing a species. Would it be better to stop using transfers for cases such as these? A solution that has been called on in the past is to demand comprehensive cost–benefit analyses and more detailed information on places that have either done IBT 'right' or know how to ameliorate these wrong outcomes based on past failures elsewhere.

10.3 Use of urban wastewater for irrigation of agricultural crops under scarcity[7]

Reuse of treated urban wastewater for beneficial purposes, including use by irrigated agriculture, can mitigate water scarcity. However, its costs and benefits are unclear. Wastewater treatment and its safe disposal are necessary requirements for urban centers to prevent possible environmental damages and health risks for humans. The reuse of treated wastewater for agricultural irrigation, as a strategy for disposal, can be beneficial to society and at the same time mitigate water scarcity (Reznik et al., 2019).

Reznik and Dinar (2021) examine the feasibility of wastewater reuse through a regional modeling framework that is applied to a real case in the Escondido region of California. Optimal results pertaining to avocado production in that region suggest significant reduction in cultivated area and preference of potable water over treated wastewater. However, supportive policies aimed at sustaining agricultural activity in the region, such as subsidizing high-quality treated wastewater for irrigation use, could be socially cost-effective.

Reuse of treated wastewater for irrigated agriculture has been on the rise worldwide (Hernandez-Sancho et al., 2015). The same trends are seen in California, where reuse of treated wastewater for different purposes has increased from 175,500 AF in 1970 to 714,000 AF in 2015. Irrigated agriculture, however, uses a decreasing share of these volumes in the last two decades. It is also reported that 417 billion gallons (nearly 1,128,000 AF) of good quality treated municipal wastewater were discharged in 2015 directly into California coastal waters (Heal the Ocean, 2018). Inland population growth in California, the distance between urban centers and agricultural regions and the vulnerability of agricultural production to California's changing climate all highlight the potential of the reuse practice and its importance for efficient management of scarce water resources in the state.

Using the Escondido case, Reznik and Dinar (2021) evaluated whether or not reuse of treated wastewater in irrigated agriculture is sustainable and economically efficient. They developed a regional model that accounts for the interaction and interdependencies among producers (urban center) and consumers (agricultural sector) of treated wastewater and the environment. The model was applied to the Escondido region in Southern California and examined the conditions under which reuse is a feasible and a sustainable alternative.

10.3.1 The Escondido region and its water challenges

The City of Escondido is located in San Diego County in southern California. The city currently supplies water to approximately 26,000 residential, commercial and agricultural customers, using surface water from both imported (approximately 80 per cent) and local sources. Due to fast development and growing population, the demands for water in the service area of the city have been growing steadily. At the same time, uncertainty of water supply availability is also growing, and the city's infrastructure capacity is becoming a binding constraint.

The city is required, by permit, to treat its sewage and dispose the treated effluents into the ocean, which is currently achieved using an ocean outfall system. According to the city projections, the range of wastewater generation in the future could be extended to the point where the existing discharge infrastructure and treatment capacity would no longer suffice in handling the volume that needs treatment and discharge.

These considerations have led Escondido to engage in several long-term planning efforts. One prominent component in such plans is the City's treated wastewater recycling activity. Currently, only a fraction of the flow of sewage to the City's wastewater treatment facility is reused for beneficial purposes. Expected insufficient capacity to meet state regulations for treated wastewater discharge to ocean outlets creates strong incentive to allocate that water locally for irrigation, in order to avoid significant expenses associated with the expansion of the ocean discharge capacity.

A possible recipient is the avocado plantations, which is the main agricultural activity in this region. Recent droughts in California have highlighted the unfavorable conditions that agriculture in this region faces. In the absence of alternative water sources to irrigate their groves, growers have to pay significantly higher rates for potable water supplied by the city. To maintain production under such conditions, growers stump trees on significant acreage, drill deep wells to access saline water and use expensive mobile desalters to avoid salinity damages. All these alternatives are costly and also affect the yield of the avocado sector.

Given these conditions and considerations, the city identified several potential solutions for treated wastewater allocation. The first, which is referred to as Non-Potable Reuse for Agriculture (NPR/Ag), develops a supply system to allocate recycled wastewater to city existing potable water consumers, specifically avocado growers. This system also includes the option to desalinate the effluents prior to its use in agriculture. The second option, named Indirect Potable Reuse (IPR), develops a separate new system to desalinate treated wastewater and convey that water to augment the city surface water supplies through storage in one of the city reservoirs.

10.3.2   Regional water and treated wastewater allocation

Reznik and Dinar (2021) developed a regional model of a general setting, relevant to many locations that is comprised of a city that needs to treat and dispose of its wastewater, an agricultural sector and the environment. The environment refers in the model to any receiving water body that can potentially be polluted by unpermitted discharge of the treated effluents. Alternatively, treated wastewater can either be discharged to a safe location (for example a remote dry-bed river or the ocean), adhering to regulation and preventing environmental damage, or be reused for beneficial purposes within the region, specifically in irrigated agriculture.

The model

The model finds an optimal regional plan of water resources allocation and infrastructure development that maximizes net benefits (or economic welfare) of all water consumers in the region, subject to several technical, hydrological and regulatory constraints. The net benefits are defined as the total economic value of water for consumption by city inhabitants and the agricultural revenue from crop output sales, minus operating and maintenance costs of the entire water system, the amortized costs of investments in infrastructure development, and monetary penalties for unpermitted discharge of treated wastewater to the environment. The model addresses farmers' heterogeneity, and includes uncertainty in key variables of both farm-level and regional decision-making processes. It therefore captures the ability of the agricultural sector to sustain changing conditions in terms of available water sources and their quality.

Data and calibration

Data was collected from public records and stakeholders in the region. Some of the components of the analytical framework were adapted to the available data.

Water supply to Escondido is provided mainly from imported surface sources. Total water supply ranges between 20,000 to 28,000 AF per year (AFY). Water availability to the region faces variability and uncertainty. To address variability, Reznik and Dinar (2021) therefore assigned a discrete probability for high- and low-water availability events.

The existing and planned water and wastewater systems in the region are represented in the model by infrastructural capacities, capital costs and operation and maintenance (O&M) costs (Table 10.3). For capital costs they use amortized values of predicted investments needed for development of alternative infrastructural components of the water system. The cost of expanding the capacity of the treatment plant is $120 per AF.

Differences in agricultural productivity and costs of avocado production in the region are captured in the model through non-linear functions that translate the use of land and water inputs to profits at the individual farm-grove level. The calibration process of these functions required high-resolution analysis of soil structure and weather in the region. In order to calibrate the parameters in these functions, Reznik and Dinar (2021) collected micro-data from individual avocado growers in the region. The growers surveyed and represented in the model are the only group of existing agricultural water consumers that will be connected (in the short term) to the new treated wastewater supply system, once in place.

Table 10.3 Escondido's water system parameters

Component description	Existing capacity (AFY)	O&M costs ($ per AF)	Capital costs, amortized values ($ per AF)
Potable water supply	Bounded by water availability	1,250	0
Wastewater collection	Not constrained	1,450	0
Treated wastewater conveyance to agriculture	0	43.99	907
Desalination of treated wastewater for agriculture	0	484.5	1,217
Desalination and conveyance of treated wastewater to augment regional supply	0	798.93	2,120
Wastewater safe disposal	24,000	29	990

Source: Reports prepared by and for the City of Escondido for the years 2012, 2014, 2017 (available by the author upon request); Reznik and Dinar (2021).

Using historical data, a certain distribution function (see Reznik and Dinar, 2021 for details) was fitted to represent the stochastic behavior of avocado prices and of precipitation in the region. In addition, Reznik and Dinar (2021) assigned discrete probabilities to different salinity and chloride levels in each of the water sources included in the model.

10.3.3 Results and discussion

The base scenario is constructed to best represent the prevailing conditions in the Escondido region as described in the previous sections. In this scenario, the ocean outlet and unpermitted discharge to the environment are the only existing alternatives for treated wastewater disposal. An optimal decision then must be made with respect to the construction of the other alternatives for treated wastewater discharge, and the allocation of all water resources available.

In terms of water quantities, results of the optimal plan suggest discharging nearly 75 per cent of the treated wastewater to the ocean, diverting nearly 15 per cent to the reservoir to augment regional potable water supply and discharging the rest to the environment (with penalty). The total volume of wastewater treated and allocated is about 13,700 AFY. Consequently, the agricultural sector in the region relies exclusively on potable water from the city. The volume of potable water diverted to avocado production is nearly 250 AFY, which is only a quarter of the actual water use by growers, as was found in the survey. Total freshwater use in the region is about 16,600 AFY, such that most of the water is consumed in the urban sector.

The implications to the agricultural sector are profound. According to the base scenario results, the land area cultivated by avocado growers shrinks, with introduction of wastewater, to less than half of the existing enterprise. Water use for that smaller farming land is also considerably lower than its observed level. Yield per cultivated acre remains mostly unchanged, and this is mainly due to use of higher quality of water allocated to agriculture. The efficient economic price of water calculated based on these outcomes is $1,500 per AF, and is higher by 40 per cent than the actual potable water price paid by farmers in the region.

Supportive agricultural policies

Considering the profound impacts on agricultural activity and the strong trade-offs referred to earlier, Reznik and Dinar (2021) examined the cost (social loss of regional economic welfare) of new supportive policies for the avocado industry. For that purpose, they designed several scenarios that differ in the water sources agriculture is allowed to use, and in the types of supportive policies. Results from the scenarios differ substantially in terms of infrastructure development and allocation of water from the various sources. However, the cost of implementing these policies under different assumptions regarding water sources utilized in agriculture are similar and amount to roughly $2 million. In relative terms, this figure is only 1.3 percent of the annual expenditures of the Escondido Water & Sewer Department. It is implied that maintaining agricultural activity in the region at its current size is warranted if this level of welfare loss is surpassed by the indirect added value from agriculture to the region (which was not quantified in the analysis).

Such indirect benefits are significant. They include job creation in avocado operations and related services, sales taxes to the City of Escondido, economic multiplier effects of avocado operations through agricultural service businesses flowing through the Escondido economy, and carbon sequestration services. And finally, in the longer run, growers may replant groves with root stocks that are more salt tolerant, allowing the City of Escondido to deliver water higher in salts/chloride, reducing their reverse osmosis expenses and charging lower price for the water.

10.3.4 Concluding remarks

The analysis of the Escondido region in southern California reveals that existing local conditions generally do not encourage reuse of treated wastewater in agriculture. Instead, the results suggest that safely discharging most of the effluent to the ocean, using existing infrastructure and augmenting regional surface water supplies with the remaining portion of wastewater after desalination is, in most cases, the preferred strategy. The analysis finds, in addition, that uncertainty, specifically with respect to water quality and its variation, as well as imposed regulatory constraints, are important drivers of this outcome.

The agricultural sector (avocados and other salinity-sensitive crops) in this region is limited in adaptation capacity, and therefore its sustainability is highly susceptible to changes in exogenous conditions. Findings suggest that supportive policies that could be crucial for the survival of the agricultural sector in the region are socially inexpensive. Therefore, implementing such policies, through capacity development and allocation of treated wastewater directly as well as after desalination, keeping agricultural water prices low and their quality high, would be warranted if benefits accrued in the region due to preservation of the agricultural sector surpass $2 million.

The empirical analysis of Escondido is tailored to the specific conditions in this region. Such conditions include the lack of inter-temporal groundwater storage, altitude differences between storage reservoirs for surface water and planned facilities for treated wastewater supply, and the distance to safe disposal of treated wastewater in the ocean. Applying the regional framework presented herein to other regions across California and across the world would be a useful extension of this example.

10.4 Summary

This chapter reviewed several emerging topics that have recently being added to the economic and policy public discourse – managed aquifer recharge, inter- and intra-basin water transfer, and reuse of municipal wastewater in irrigated agriculture. All three topics are part of an effort to ease the level of scarcity of water resources, while keeping the profitability and sustainability of water-using sectors at reasonable levels. All these three emerging topics are seen in many parts of the world, and thus they are of general interest beyond a given locality and were selected to be shared with the readers of this book.

Notes

1. This section is based on work conducted with Ami Reznik during his postdoc stay at the School of Public Policy, University of California, Riverside, and published as Reznik et al. (2020).
2. A social planner solution corresponds to the optimization problem where all decision and state variables of the system are determined to maximize the

present value of net gains of the entire region, ignoring income distribution implications.

3. Deficit irrigation is a technique where irrigators stress the crop during certain growing stages where minimal damage is expected.
4. This section is based on work conducted with Logan Purvis during graduate study at the School of Public Policy, University of California, Riverside and published as Purvis and Dinar (2019).
5. Stress = 1,000–1,700; scarcity = 500–1,000; severe scarcity ≤ 500 m^3 per person per year.
6. We should emphasize that the five attributes are not mutually exclusive. This is especially important to mention in the case of ER, EE and EJ, all somehow related to environmental aspects. However, they do not overlap, but rather each of them relates to specific and particular effects that have been seen separately in water-related negative externalities.
7. This section is based on work conducted with Ami Reznik during his postdoc stay at the School of Public Policy, University of California, Riverside and published as Reznik and Dinar (2021).

References

Boardman, A. E., D. H. Greenberg, A. R. Vining, and D. L. Weimer, 2018. *Cost–Benefit Analysis: Concepts and Practice*, 5th ed. Cambridge: Cambridge University Press.

Dinar, A. and R.M. Saleth, 2005. Can water institution be cured: A water institution's health index, *Water Science and Technology: Water Supply Journal*, 5(6):17–40.

Dudley, N. J., and W. F. Musgrave, 1988. Capacity sharing of water reservoirs. *Water Resources Research*, 24(5):649–58.

Gupta, J. and P. van der Zaag, 2008. Interbasin water transfers and integrated water resources management: Where engineering, science and politics interlock. *Physics and Chemistry of Earth*, Parts A/B/C 33(1):28–40.

Heal the Ocean, 2018. Inventory of municipal wastewater discharges to California coastal waters. Accessed 20 March, 2022. https://www.wastewater-inventory.healtheocean.org/.

Hernandez-Sancho, F., B. Lamizana-Diallo, J. Mateo-Sagasta, and M. Qadeer, 2015. *Economic Valuation of Wastewater – The Cost of Action and the Cost of No Action*. Nairobi: United Nations Environment Programme.

Jagannathan, N. V., M. A. Shawky, and A. Kremer, 2009. *Water in the Arab World: Management Perspectives and Innovations*. Washington, DC: World Bank.

Mahabaleshwara, H. and H. M. Nagabhushan, 2014. Inter-basin water transfers in India – a solution to hydrological extremities. *International Journal of Resource Engineering and Technology*, 3(3):530–7.

Purvis, L. and A. Dinar, 2019. Are intra- and inter-basin water transfers a sustainable policy intervention for addressing water scarcity? *Water Security*, 9:100058.

Reznik, A. and A. Dinar, 2021. Local conditions and the economic feasibility of urban wastewater recycling in irrigated agriculture: Lessons from a stochastic regional analysis in California. *Applied Economic Perspectives and Policy*, https://doi.org/10.1002/aepp.13198.

Reznik, A., A. Dinar, S. Bresney, Laura Forni, Brian Joyce, Steven Wallander, Daniel Bigelow, and Iddo Kan, 2020. *Managed aquifer recharge as a strategy to mitigate drought impacts in irrigated agriculture: Role of institutions and policies with application to California.* UCR SPP Working Paper Series, March, 2020 WP# 20-04. Accessed 13 July, 2022. https://spp.ucr.edu/sites/g/files/rcwecm1611/files/2020-11/102820_%20Reznik%20et%20al%20MAR%20WP%20Text%20and%20Figures_0.pdf.

Reznik, A., A. Dinar, S. Bresney, L. Forni, B. Joyce, S. Wallander, D. Bigelow, and I. Kan, 2021. Can managed aquifer recharge mitigate drought impacts on California's irrigated agriculture? The role for institutions and policies. *ARE Update*, 24(4): 5–8.

Reznik, A., A. Dinar, and F. Hernández-Sancho, 2019. Treated wastewater reuse: An efficient and sustainable solution for water resource scarcity. *Environmental and Resource Economics*, 74(4):1647–85.

Rinaudo, J. D. and B. Barraque, 2015. Inter-basin transfers as a supply option: the end of an era? In Q. Grafton, K.A. Daniell, C. Nauges, J.-D. Rinaudo, N.W.W. Chan (eds.), *Understanding and Managing Urban Water in Transition*, 175–200. Berlin: Springer.

Snaddon, C. D., B.R. Davies, and M. J. Wishart, 1999. A global overview of inter-basin water transfer schemes, with an appraisal of their ecological, socio-economic and sociopolitical implications, and recommendations for their management. Report TT120/00, December. Pretoria, South Africa: Water Research Commission.

UNESCO, 1999. Interbasin water transfer: Proceedings of the international workshop, UNESCO, Paris, 25–27 April 1999. Accessed 19 March, 2022. https://unesdoc.unesco.org/ark:/48223/pf0000161070.

WWAP (UNESCO World Water Assessment Programme), 2019. *The United Nations World Water Development Report: Leaving No One Behind.* Paris: UNESCO, 2019.

Yevjevich, V. S., 2001. Water diversions and inter-basin transfers. *Water International*, 26(3):342–8.

Zhuang, W., 2016. Eco-environmental impact of inter-basin water transfer projects: A review. *Environmental Science Pollution Research*, 23(13):12867–79.

11   Summary and concluding remarks to *Advanced Introduction to Water Economics and Policy*

This book presents an attempt to look at water management from the prism of economic perspectives. In this book I attempted to cover, in a non-technical way, different aspects of the problems that economic approaches attempt to address. The book covers various important water-using sectors, such as agriculture (Chapter 3) and residential (Chapter 4) sectors; different types of water, such as surface water, groundwater (Chapter 6) and wastewater; and different scales and scopes of water use, such as local, farm-level, regional-level and international-level uses (Chapter 8).

Each of the sectors, types of water used and scales of use necessitate different economic considerations, models and policy interventions. We have seen how in the case of groundwater there is a need for use of models that take into account the interaction with surface water; include considerations of pollution from both irrigated agriculture pollution and residential wastewater disposal; and compare policy implications for different groups in the economy.

As we well know, economic considerations become relevant as the good or resource becomes scarcer. Scarcity of water is determined by the physical amount of water available for use, or by its distribution as affected by climate change (Chapter 9), but also by its quality (Chapter 7). Water of lower quality is less suitable for various uses (e.g., nitrates in drinking water, salinity in irrigation water). These are two aspects of water management that have to be addressed, using economic principles.

We also realized that the economic toolkits at our disposal are very rich. They include pricing interventions that can send signals of the scarcity value of the water resources available in a given region or state. Regarding pricing water consumption, we learned that different pricing scheme types may work more effectively in the agricultural sector and others in the residential sector due to differences in precision and monitoring abilities of the regulatory agencies.

The quota tools may work more effectively in the agricultural sector than in the residential sector since the demand for water in agriculture is more seasonally distributed and in the residential sector it is more or less constant over the year.

Subsidies to promote purchasing more efficient equipment for water use (irrigation technologies in the agricultural sector and shower heads in the residential sector) have been proven effective and able to help reach targets of water conservation in these sectors.

While we reviewed and discussed use of different economic tools for managing sectoral water and water of different sources, a comparison of the relative effectiveness of these tools may indicate different levels of effectiveness and priorities, depending on the specific conditions (physical, social, institutional) that exist in many locations around the world. Therefore, it is quite important to be careful when recommending different economic tools to address water management under different conditions.

Two additional chapters in this book addressed two emerging aspects that have gained importance in the literature and in actual situations: management of international water (Chapter 8) and effects of climate change on the water economy (Chapter 9). We also introduced three emerging aspects of water management that are used to ease water scarcity (Chapter 10) – managed aquifer recharge, inter- and intra-basin water transfers, and reuse of treated wastewater in irrigated agriculture.

As water scarcity becomes a limiting factor in many parts of the world, the management of water that is shared by several riparian states is a critical aspect of sustainability of regions and relations between states. Economic analysis of international water management includes concepts such as social planner solution (allocation) that have not been used in economic analyses of single decision-makers (such as a farmer that needs to allocate scarce water across fields or crops). In addition, as we realized in this book, management of international water necessitates the introduction of political considerations into the analytical framework we develop. This addition is important to refer to and important to recognize since interactions between decisions in cases of international river basins and the performance of the domestic water sector become more and more interconnected.

Another important aspect of water economics that was addressed in the book is the environment-water interactions and management (Chapter 5). Recognition of the environmental services of water resources has been recognized in recent works and found to be important and significant. We discussed several ways that environmental health can remain sustainable for the sake of the economy, and various policy interventions that could be used.

We also paid attention in this book to climate change impacts on the economy via its impact on the water sector both in a special dedicated chapter (Chapter 9) and across most other chapters via dealing with climate change-induced water scarcity. There is evidence that climate change would affect the water sector in many ways that will impact the performance of the water-consuming sectors. Changes to the annual mean precipitation, changes to the distribution of precipitation across the year, changes to temperature and change to humidity all alter the water system performance and affect the ability of water users to properly utilize that resource. We discussed the various ways governments, water utilities and individual users (farmers, households) can respond to climate-affected water systems to allow minimum damage. The objective of these adaptation measures is to minimize the damage from climate change. Damage would persist with adaptation, but at a lower level. We also learned that damage from climate change would affect different nations and groups of users to a different extent. We have not addressed the inequality impacts of the climate change, but it is a major concern for policy makers and should be addressed.

11.1   What the book has not covered

Naturally, the book could not cover all issues related to water economics. Some of the issues that we addressed were also not addressed in depth. Further reading is listed and explained in section 11.2 below.

Two important aspects of economic analysis were not addressed by this book. First, the strategic behavior of decision-makers at various levels was not covered, though it is important for the understanding of water users' decision-making processes, especially under situations of water scarcity. The reader will be referred below to supplemental reading on the role of game theory in water resources. Another important aspect of economic analysis of water resources is the handling of the water sector as part of the entire economy and not as a stand-alone sector. As we already well know, policies implemented in various sectors (e.g., international trade, labor) will have direct and indirect effects on water-consuming sectors. The reader will be referred to work on economy-wide analysis of water resources in the next section. And finally, this book has not covered the important aspect of the 'conjunctive use' of different types of water (e.g., surface water, groundwater, wastewater, brackish water). Using waters of different types conjunctively could be a very attractive solution to water scarcity. But there are many tradeoffs, such as water quality consequences, third party effects and infrastructure adjustment considerations, all of which must be entered into the equation. All these aspects were not covered in the book and are provided below as supplementary material to interested readers.

11.2   Suggestions for further reading

In the following I provide three suggested readings for important aspects that have not been covered in this book.

11.2.1   Conjunctive use

Conjunctive use of various types of water are analyzed with many examples in Coe (1990). The work focuses on two types of water – surface water and groundwater – but could easily be extended to others. Conjunctive use could benefit the user facing water scarcity at lower costs than when each source and the infrastructure associated with its operation (dams and reservoirs) are operated separately. Physical, operational, financial and institutional constraints and difficulties could be encountered by the conjunctive project. The economic analysis role is to identify these difficulties and quantify them in terms of costs and benefits.

11.2.2 Water and economy-wide considerations

Somewhat similar is the conjunction of the entire economy when making water-related decisions. Dinar (2014) discusses a framework that was developed and applied in several countries regarding various policy interventions aimed at improving water allocation decisions with an economy-wide context. It is based on the 'macro–micro linkage' framework that facilitates assessment of various linkages among policies and their impacts within different connected individual sectors and the economy. Drawing on studies in Mexico, Morocco, South Africa and Turkey, the analysis in Dinar (2014) finds difficult trade-offs among various water and non-water policy objectives, including priorities placed on different sectors, regional advantages and general economic efficiency gains versus broader social impacts. The comparison of policy impacts suggests how policy makers may use such information to rank the policy interventions according to the emphasis placed on their objectives. Certainly solutions to water problems obtained in an economy-wide framework differ from solutions obtained at sectoral or local levels, and this difference is critical for policy purposes.

11.2.3 Water and strategic behavior

Water resources face new challenges such as scarcity, growing populations and massive development. These have led to increased competition over water resources and subsequent elevated pollution levels. In addition, climate change is expected to exacerbate the situation and affect the hydrological cycle, leading to increased variability in water supplies across time and space and adding uncertainty to water allocation decisions. Future investments in water resource projects will be substantial, needing much more stable rules for cost allocations among participating users and over time. The nature of water disputes among users may vary from local to regional, state and international levels. All of the above suggests that strategic behavior models are needed. Dinar and Hogarth (2015) argue that while game theory approaches and applications to water resources have advanced over the years, much more is required to make different game theory models relevant to the new challenges facing the water sector. Dinar and Hogarth (2015) review the main contributions of game theory in water resources over the past 70 years, comparing the issues faced by water resources and those that the sector is most likely to face in the coming future. The work they present covers various sectors, various types of water and various scopes (local, regional and international).

References

Coe, J. J., 1990. Conjunctive use—advantages, constraints, and examples. *Journal of Irrigation and Drainage Engineering*, 116(3):427–43.

Dinar, A., 2014. Water and economy-wide policy interventions. *Foundations and Trends in Microeconomics*, 10(2):85–164.

Dinar, A. and M. Hogarth, 2015. Game theory and water resources: Critical review of its contributions, progress and remaining challenges. *Foundations & Trends in Microeconomics*, 11(1–2):1–139.

Index

on energy used for pumping 51
water 23, 53
on water extractions 50–51
TDP *see* tradeable discharge permits (TDP)
technical efficiency of irrigation water use 17
technological ability 6
technology adoption 3, 52
technology standards 60
Tellez-Foster, E. 51
thermoelectric power production 11
Tickner, D. 37
tradeable discharge permits (TDP) 63–4
tradeable water rights 72
traditional regulatory intervention 26
transboundary water 67
treaties 69
Tsvetanov, T. 27

UNCNRET *see* United Nations Centre for Natural Resources, Energy and Transport (UNCNRET)
UNESCO 78
unilateral negative externalities 70
United Nations Centre for Natural Resources, Energy and Transport (UNCNRET) 68
United States, population and water use in 10–12
UN-Water 78
upstream 70
urban wastewater for irrigation of agricultural crops 102–9
urban water utilities 34

Ward M. H. 48
wastewater 49
allocation 104, 105–7
disposal 107, 112
facility 83
production 83
reuse 102
sector and agricultural sector 82–5
treatment

costs, climate change on 82–4
facility 104
plants 83
sectors 82
water allocations 81
to address environmental flows 39–42
concepts 75
water and economy-wide considerations 116
water and strategic behavior 116–17
water availability and use 5–6
changes in global water use 7–10
total available renewable natural water resources in world 7
trends in water use in the United States 10–12
water bodies 16, 38, 59, 63, 70, 80, 105
water budget rate (WBR) 28, 29
water collection systems 82
Water Conservation Act 29
water conservation policies in California's residential sector 29–34
water conservation regulations in California 33
water conserving fixtures and landscape, rebate programs for 27–8
water-consuming economic activities 95
water-consuming sectors 114, 115
water consumption 97, 113
household 25–9
in residential sector 26–7
water-dependent ecosystems 17, 36, 37
water economics and policy
inter-basin and intra-basin water transfers 95–102
managed aquifer recharge (MAR) 88–95
urban wastewater for irrigation of agricultural crops under scarcity 102–9
water externalities 102
water extractions
caps on 49–50